Louis Nirenberg (Ed.)

Pseudo-differential Operators

Lectures given at a Summer School of the
Centro Internazionale Matematico Estivo (C.I.M.E.),
held in Stresa (Varese), Italy,
August 26-September 3, 1968

Springer

FONDAZIONE CIME ROBERTO CONTI

C.I.M.E. Foundation
c/o Dipartimento di Matematica "U. Dini"
Viale margagni n. 67/a
50134 Firenze
Italy
cime@math.unifi.it

ISBN 978-3-642-11073-3 e-ISBN: 978-3-642-11074-0
DOI:10.1007/978-3-642-11074-0
Springer Heidelberg Dordrecht London New York

Reprint of the 1st ed. C.I.M.E., Ed. Cremonese, Roma 1969
With kind permission of C.I.M.E.

Printed on acid-free paper

Springer.com

CENTRO INTERNAZIONALE MATEMATICO ESTIVO
(C.I.M.E.)

2° Ciclo - Stresa dal 26 Agosto al 3 Settembre 1968

"PSEUDO-DIFFERENTIAL OPERATORS"

Coordinatore: L. Nirenberg

S. AGMON	: Asymptotic formulas with remainder estimates for eingevalues of elliptic operators	pag. 1
J. BOKOBZA-HAGGIAG	: Une définition globale des opérateurs pseudo-différentiels sur une variété différentiable	pag. 11
L. BOUTET DE MONVEL	: Pseudo-differential operators and analytic functions	pag. 37
A. CALDERON	: A priori estimates for singular integral operators	pag. 85
O. CORDES	: Testo non pervenuto	
F.B. JONES	: Characterization of spaces of Bessel potentials related to the heat equation	pag. 143
J.J. KOHN	: Pseudo-differential operators and non-elliptic problems	pag. 157
R. SEELEY	: Topics in pseudo-differential operators	pag. 167
E. SHAMIR	: Boundary value problems for elliptic convolutions systems	pag. 307
I.M. SINGER	: Elliptic operators on manifolds	pag. 333

CENTRO INTERNAZIONALE MATEMATICO ESTIVO

(C.I.M.E.)

S. AGMON

ASYMPTOTIC FORMULAS WITH REMAINDER ESTIMATES FOR EINGEVALUES OF ELLIPTIC OPERATORS

Corso tenuto a Stresa dal 26 Agosto al 3 Settembre 1968

ASYMPTOTIC FORMULAS WITH REMAINDER ESTIMATES FOR EINGEVALUES OF ELLIPTIC OPERATORS

by

S. Agmon (Hebrew University)

We propose to discuss in this lecture a number of results related to the problem of eigenvalue distribution of elliptic operators. We start with some classical results. Let Δ be the Laplacian in R^n and consider the eigenvalue problem:

$$(1) \qquad -\Delta u = \lambda u \quad \text{in } \Omega, \qquad u = 0 \quad \text{on } \partial\Omega,$$

where Ω is a bounded open set in R^n. Let $\{\lambda_j\}$ be the sequence of eigenvalues of (1), each repeated according to its multiplicity and set

$$(2) \qquad N(t) = \sum_{\lambda_j < t} 1.$$

Then according to a wellknown theorem of Weyl one has:

$$(3) \qquad N(t) = \gamma t^{n/2} + o(t^{n/2}), \quad t \to +\infty,$$

γ some constant. The formula (3) was further improved by Courant who gave following estimate to the remainder term:

$$(4) \qquad N(t) - \gamma t^{n/2} = O(t^{(n-1)/2} \log t).$$

For the corresmonding case of the Laplacian on a compact manifold Avakumovic [4] proved a somewhat better result that (4), namely that

$$(5) \qquad N(t) - \gamma t^{n/2} = O(t^{(n-1)/2}).$$

The remainder estimate (5) is actually the best possible estimate which one can expect in general. This, as was observed by Avakumovic [4] (see also [12]), follows by considering the example of the Laplacian Δ_S on the sphere S^n. In this case the different eigenvalues are $\mu_j = j(j+n-1)$, each μ_j having a multiplicity:

$$\binom{n+j}{n} - \binom{n+j-2}{n}$$

It then follows that

(6) $$N(\mu_j + o) - N(\mu_j - o) \geq c\,\mu_j^{(n-1)/2}$$

for some constant $c > o$. From (6) one concludes that (5) cannot be improved for the Laplacian on the sphere. The same example also shows that a more refined asymptotic formula for $N(t)$ with a second term in the asymptotic formula need not exist.

There is one case where (5) can be improved. This is the case of the Laplace operator on the torus R^n mod 2π. In this case $N(t^2)$ is nothing else but the number of lattice points inside the sphere of radius t (center at the origin). It follows from classical results (e.g. [13]) that

(7) $$N(t) - \gamma t^{n/2} = 0(t^{(n/2)-n/(n+1)}) .$$

Returning to the question of eigenvalue distribution for the Dirichlet problem (1) (or any other self-adjoint differential boundary value problem for Δ) it seems very plausible that the optimal remainder estimate (5) holds also in this case. Up to now, however, this result was established only in the very special case when Ω is a polyhedron in 2 or 3 dimensions ([5] and [8;9]) .

We consider now the case of a general elliptic operator. (For simplicity we shall consider the case of a single differenttial operator acting on functions defined on an open set of R^n. Most of the results, however, admit generalizations to elliptic systems defined on manifolds) . Let A be a formally self-adjoint (positive) differential operator of order m in $\Omega \subset R^n$ (C^∞ coefficients) . Let $\tilde{A}$ be a self-adjoint realization of A in the Hilbert space $L_2(\Omega)$. Assume also that $\tilde{A}$ is bounded from below and that it has a discrete spectrum consisting of eigenvalues of finite multiplicity. As before let N(t) denote the number of eigenvalues < t. The following general result on the asymptotic behavior of N(t) holds.

Theorem 1. Suppose that Ω is a bounded open set possessing the cone property. Suppose furthermore that for some integer $k > n/m$ the domain of definition of $\tilde{A}^k$ is contained in $H_{km}(\Omega)$. Then

(8) $$N(t) = \gamma t^{n/m} + o(t^{n/m}), \qquad t \longrightarrow \infty,$$

$$\gamma = \int_{\Omega}\int_{A'(x,\xi) < 1} d\xi dx .$$

Theorem 1 yields the main term in the asymptotic eigenvalue distribution formula for a wide class of elliptic boundary value problems (e.g. [1], [6] and references given there). Recently the theorem was improved to yield an estimate for the remainder in (8). The following result holds [2].

Theorem 2. Under essentially the same conditions as those in Theorem 1:

(9) $$N(t) - \gamma t^{n/m} = O(t^{(n-\sigma)/m}), \quad t \longrightarrow \infty ,$$

for any $\sigma < 1/2$ in the general case and any $\sigma < 1$ if the principal part A' has constant coefficients.

The basic step in proving Theorem 1 or Theorem 2 is the derivation of the corresponding asymptotic formula for the spectral function:

(10) $$e(t;x,x) = \sum_{\lambda_j < t} \varphi_j(x)\, \overline{\varphi_j(y)} ,$$

where $\{\varphi_j(x)\}$ is the normalized sequence of eigenfunctions. Thus in order to prove Theorem 1 one uses the asymptotic formula (given by Carleman [7] for second order operators and by Gårding [10] in the general case):

(11) $$e(t;x,x) = c(x)t^{n/m} + o(t^{n/m}), \quad t \to \infty, \quad c(x) = \int_{A'(x,\xi)<1} d\xi$$

By integration of (11) over Ω one arrives at the asymptotic formula (8). (Actually one needs additional estimates for the spectral function in order to justify the integration step since (11) is uniform in x only on compact subsets of Ω). In analogy the proof of Theorem 2 is based on the following estimate for the remainder in (11) established recently by Agmon and Kannai [3] and Hörmander [11]:

(12) $$e(t;x,x)-c(x)t^{n/m} = 0(t^{(n-\sigma)/m})$$

for any $\sigma < 1/2$ in the general case and any $\sigma < 1$ if the principal part A' has constant coefficients.[1)]

We give now some indications about the derivation of (12). Without loss of generality we shall assume that $m > n$ (this could always be achieved replacing $\tilde{A}$ by some power $\tilde{A}^k$), and that $\tilde{A}$ is positive. Let $R_\lambda = (\tilde{A}-\lambda)^{-1}$ be the resolvent operator. Because $m > n$ it follows that R_λ is an integral with a continuous kernel $R_\lambda(x,y)$. One has the relation:

(13) $$R_\lambda(x,x) = \int_0^\infty (t-\lambda)^{-1}\, de(t;x,x) .$$

The main idea which goes back to Carleman is to determine the asymptotic behavior of $R_\lambda(x,x)$ in the complex λ-plane and then to deduce from this information, via the relation (13), the asymptotic behavior of the spectral function. It should be noted, however, that for the derivation of remainder

---

1) It should be noted that the various asymptotic results on spectral functions actually hold in the general situation when a spectral function exists. i.e. for any semi-bounded from below self-adjoint realization of A (no assumptions on the spectrum or on Ω).

estimates such as (12) one needs information on the asymptotic behavior of the resolvent kernel $R_\lambda(x,x)$ near the spectrum of $\widetilde{A}$. Thus the crucial result proved in [3] is the following.

<u>Theorem 3.</u> The resolvent kernel $R_\lambda(x,x)$ admits an asymptotic expansion of the form:

$$R_\lambda(x,x) \sim (-\lambda)^{n/m-1} \sum_{j=0}^{\infty} c_j(x)(-\lambda)^{-j/m} \quad . \tag{14}$$

valid for $\operatorname{Re}\lambda > 0$, $|\operatorname{Im}\lambda| \geq |\lambda|^{(m-\theta)/m+\varepsilon}$, $|\lambda| \geqslant 1$ (and also for $\operatorname{Re}\lambda \leq 0$, $|\lambda| \geq 1$) where $\theta = 1/2$ in the general case, $\theta = 1$ if A' has constant coefficients, and ε is an arbitrary positive number. Here $c_j(x)$ are certain C^∞ functions depending only on the differential operator A.

The deduction of Theorem 2 from Theorem 3 can be achieved with the following semi-inversion formula of (13) due to Pleijel:

$$\left| e(t;x,x) - \frac{1}{2\pi i}\int_\Gamma R_\lambda(x,x)d\lambda \right| \cdot 2\tau |R_\zeta(x,x)| \tag{15}$$

where $\zeta = t + i\tau$ with $\tau > 0$ and Γ is any contour from $\bar{\zeta}$ to ζ not intersecting the positive axis. By choosing for Γ a contour of the form: $|\operatorname{Im}\lambda| = |\lambda|^{(m-\theta)/m+\varepsilon}$, $0 \leqslant \operatorname{Re}\lambda \leqslant t$, using (15) and (14), one arrives easily at the remainder estimate (13).

In order to prove Theorem 3 one needs to construct a good parametrix for the resolvent kernel which will yield also a good approximation near the spectrum. This can be done for instance with the aid of the calculus of pseudo-differential operators (this procedure was followed in [11] but not in [3] where a somewhat different method was used).

Finally we mention that very recently, using new classes of pseudo-differential operators, Hörmander [12] succeeded in proving the following optimal remainder estimate for the spectral function:

$$e(t;x,x)-c(x)t^{n/m} = 0(t^{(n-1)/m}) \ . \tag{16}$$

With the aid of this result it is possible to improve now Theorem 1 in the case of a single operator by showing that in the general case

$$N(t)-\gamma t^{n/m} = 0(t^{(n-1)/m+\varepsilon}) \tag{17}$$

for any $\varepsilon > o$. It would be interesting to know whether (17) holds with $\varepsilon = o$. As was remarked before it is not known whether this is true even in the case of the Dirichlet boundary value problem for the Laplace operator.

S. Agmon

REFERENCES

[1] Agmon, S., On kernels, eigenvalues, and eigenfunctions of operators related to elliptic problems, Comm. Pure Appl. Math. 18 (1965), 627-663.

[2] Agmon, S., Asymptotic formulas with remainder estimates for eigenvalues of elliptic operators, Arch. Rat. Mech. Anal. 28 (1968), 165-183.

[3] Agmon, S. and Y. Kannai, On the asymptotic behavior of spectral functions and resolvent kernels of elliptic operators, Israel J. Math. 5 (1967), 1-30.

[4] Avakumovic, V. G., Ueber die Eigenfunktionen auf geschlossenen Riemannschen Mannigfaltigkeiten, Math. Z. 65 (1956), 327-344.

[5] Bailey, P. B. and F. H. Brownell, Removal of the log factor in the asymptotic estimates of polygonal membrane eigenvalues, J. Math. Appl. 4(1962), 212-239.

[6] Browder, F. E., Asymptotic distribution of eigenvalues, and eigen-functions for non-local elliptic boundary value problems I., Amer. J. Math. 87(1965), 175-195.

[7] Carleman, T., Propriétés asymptotiques des fonctions fondamentales des membranes vibrantes, C.R. du 8ème Congrès de Math. Scand. Stockholm 1934 (Lund 1935), 34-44.

[8] Fedosov, B. V. Asymptotic Formulas for the eigenvalues of the Laplace operator in the case of a polygonal domain, Dokl. Akad. Nauk SSSR 151 (1963), 786-789.

[9] Fedosov, B. V. Asymptotic formulas for the eigenvalues of the Laplace operator for a polyhedron, Dokl. Akad. Nauk SSSR 157 (1964), 536-538.

[10] Gårding, L., On the asymptotic properties of the spectral function belonging to a self-adjoint semi-bounded extension of an elliptic differential operator, Kungl. Fysiogr. Sällsk. i Lund Forth. 24 (1954), 1-18.

[11] Hörmander, L., On the Riesz means of spectral functions and eigenfunction expansions for elliptic differential operators. To appear.

[12] Hörmander, L., The spectral function of an elliptic operator, To appear.

[13] Landau, E. Einführung in die Zahlentheorie II, Leipzig 1927.

CENTRO INTERNAZIONALE MATEMATICO ESTIVO

(C.I.M.E.)

J. BOKOBZA-HAGGIAG

UNE DEFINITION GLOBALE DES OPERATEURS PSEUDO-DIFFERENTIELS SUR UNE VARIETE DIFFERENTIABLE

Corso tenuto a Stresa dal 26 Agosto al 3 Settembre 1968

UNE DEFINITION GLOBALE DES OPERATEURS PSEUDO-DIFFERENTIELS SUR UNE VARIETE DIFFERENTIABLE.

par

Juliane Bokobza-Haggiag

(Purdue University)

Introduction.

Nous introduisons dans ce qui suit une définition globale des opérateurs pseudo-différentiels sur une variété différentiable et un calcul symbolique qui permet d'établir une correspondance linéaire bijective entre les opérateurs pseudo-différentiels modulo les opérateurs régularisants d'une part et une classe de symboles modulo les symboles qui sont à décroissance rapide sur les fibres de l'espace cotangent d'autre part.

L'idée de ce calcul est basée sur le fait que la formule

$$(A \varphi)(x) = \int f(x, \xi) d\xi \int e^{-2i\pi(y-x).\xi} \varphi(y) dy$$

qui définit un opérateur pseudo-differentiel sur $\mathbb{R}^n$, si f a certaines propriétés de régularité et de croissance à l'infini, prend un sens sur une variété si l'on y remplace $y-x$ par un vecteur tangent en x à la variété, soit $v(x,y)$, "infinitésimalement égal" à $y-x$, et si l'on prend quelques précautions supplémentaires destinées à faire converger l'intégrale et à lui assurer un sens intrinsèque .

J. Bokobza

CHAPITRE I

LINEARISATION D'UNE VARIETE DIFFERENTIABLE.

Définition (I. 1) . Soient X une variété réelle de classe C^{∞}, paracompacte et $T(X)$ son espace tangent. Soit v une application de classe C^{∞} de XxX dans $T(X)$ telle que le diagramme

$$\begin{array}{ccc} XxX & \xrightarrow{\ v\ } & T(X) \\ & {\scriptstyle p_1}\searrow \quad \swarrow & \\ & X & \end{array}$$

soit commutatif, p_1 désignant la première projection du produit XxX sur X. On dira que v est une linéarisation de X si les conditions (i) et (ii) sont vérifiées:

(i) pour tout $x \in X, v(x, x) = 0$;

(ii) pour tout $x \in X$, la différentielle de l'application $y \longmapsto v(x, y)$, laquelle est une application linéaire de $T_y(X)$ dans $T_x(X)$, est au point $y=x$ l'application identique de $T_x(X)$.

Signalons dès maintenant que plus que de l'application $d_y v(x, y)$ nous aurons besoin de sa transposée que nous notons l_x^y : c'est l'application linéaire de $T_x^*(X)$ dans $T_y^*(X)$ définie par

$$l_x^y(\xi) = d_y < v(x, y), \xi >$$

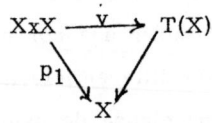

pour tout ξ appartenant à $T_x^*(X)$, où $T^*(X)$ désigne l'espace cotangent de X; la condition (ii) s'exprime bien sûr par le fait que pour tout $x \in X$ 1_x^x est l'application identique de $T_x^*(X)$.

Proposition (I. 1). Soient X et Y deux variétés réelles de classe C^∞, paracompactes, et h une immersion $Y \to X$, c'est à dire une application C^∞, de rang partout égal à la dimension de Y.

Soit $h^*(T(X))$ le fibré sur Y image réciproque par h de T(X), à un sous-fibré duquel on peut identifier T(X) (au moyen de la différentielle de h) et soit H un fibré supplémentaire de T(Y) dans $h^*(T(X))$ (tel, par exemple, qu'on peut le définir par choix d'une structure riemanienne sur $h^*(T(X))$. Soit q la projection de $h^*(T(X))$ sur T(Y) parallèlement à H.

Alors si v est une linéarisation sur X, on définit une linéarisation sur Y en posant $(h^*v)(y_1, y_2) = q(v(hy_1, hy_2))$.

En effet $d_{y_2}(h^*v)(y_1, y_2) = q \circ {}^t 1_{hy_1}^{hy_2} \, dh(h_2)$, d'où résulte immédiatement la condition (ii) puisque $1_{hy_1}^{hy_1} = I$ et que l'application q est une inverse à gauche de dh.

Notons que h^*v est définie canoniquement par la donnée de h et v lorsque h est un difféomorphisme local.

Corollaire. Toute variété admet une linéarisation.

Cela résulte en effet du théorème de Whitney et de la proposition (I. 1).

Proposition (I. 2). Soit v une linéarisation de la variété X . Il existe un voisinage ouvert Ω de Δ, diagonale de XxX, vérifiant les propriétés suivantes:

(iii) la restriction de v à Ω est un difféomorphisme de Ω sur un voisinage ouvert de la section nulle de T(X).

(iv) le deux projections de $\bar{\Omega}$ sur X sont des appli cations propres.

Un tel ouvert Ω sera appelé un domaine de la linéarisation v .

En outre, on peut supposer que Ω est symétrique et que pour tout $x \in X$, la coupe $\Omega_x = \{ y \in X$ tels que $(x, y) \in \Omega \}$ est connexe.

En effet, d'après le théorème des fonctions implicites, il existe, pour tout $x \in X$, un voisinage ouvert U_x de x tel que la restriction de v à $U_x x U_x$ soit un difféomorphisme. On peut bien sûr , supposer U_x relativement compact.

X étant paracompacte, on sait que l'on peut définir un recouvrement ouvert localement fini (V_λ) de X tel que si $V_\lambda \cap V_\mu \neq \emptyset$ il existe $\in X$ tel que $V_\lambda \cup V_\mu \subset U_x$.

Alors $\Omega = \bigcup_\lambda V_\lambda \times V_\lambda$ vérifie les conditions (iii) et (iv): pour la condition (iii), il n'y a que l'injectivité à verifier; mais si $v(x, y) = v(x, z)$, où $(x, y) \in V_\lambda \times V_\lambda$ et $(x, z) \in V_\mu \times V_\mu$, alors $V_\lambda \cup V_\mu \subset U_t$ pour un certain $t \in X$, de sorte que (x, y) et (x, z) appartiennent tous deux à $U_t x U_t$, d'où $y = z$.

Par ailleurs, si K est une partie compacte de X, (x, y) ne peut appartenir à Ω , x appartenant à K , que si y appartient à la réunion des ouverts relativement compacts V_λ qui coupent K , lesquels sont en nombre fini, d'où (iv).

Pour assurer les deux dernières conditions, choisissons pour tout $x \in X$ un voisinage ouvert connexe W_x de x tel que $W_x \times W_x \subset \Omega$. Alors $\Omega' = \bigcup_{x \in X} W_x \times W_x$ répond à la question car $\Omega'_x = \bigcup_{x \in W_y} W_y$ est connexe, et Ω' est, bien sûr, symétrique. Cette démonstration prouve d'ailleurs que tout voisinage de Δ contient un domaine de v , sumétrique et à coupes connexes.

CHAPITRE II

ESPACES DE SYMBOLES ET D'OPERATEURS.

Dans tout ce qui suit, X est une variété réelle de classe C^∞ et de dimension n, que l'on suppose, pour simplifier, dénombrable à l'infini.

Définition (II. 1). Soit m un nombre réel. $C^\infty(T^*(X);m)$ est l'espace des fonctions F de classe C^∞ sur $T^*(X)$, à valeurs complexes, telles que pour tout compact K de X contenu dans un domaine de coordonnées, pour tout choix d'un système de coordonnées dans un voisinage de K , et quels que soient les multi-indices p et q, il existe une constante C telle que:

$$\left| D_x^p \, \partial_\xi^q F(x,\xi) \right| \leq C \, |\xi|^{m-|q|}$$

$$\text{pour} \quad x \in K \quad \text{et} \quad \xi \in T_x(X), \ |\xi| \geq 1$$

Remarque:

On pose ici

$$D_x^p = \frac{1}{(2i\pi)^{|p|}} \; \frac{\partial^{|p|}}{\partial x_1^{p_1} \ldots \partial x_n^{p_n}} \qquad \text{et}$$

$$\partial_\xi^q = \frac{\partial^{|q|}}{\partial \xi_1^{q_1} \ldots \partial \xi_n^{q_n}}$$

il y aura lieu d'introduire également des dérivations par rapport aux cordonnées de l'espace tangent sous la forme

$$D_\eta^p = \frac{1}{(2i\pi)^{|p|}} \; \frac{\partial^{|p|}}{\partial \eta_1^{p_1} \ldots \partial \eta_n^{p_n}} .$$

Ces dérivations ne prennent bien sûr un sens que moyennant la donnée d'un système de coordonnées locales ou celle d'une base en un point donné de l'espace tangent ou cotangent; mais toutes les formules globales qui suivront et feront intervenir de telles dérivations ont un sens intrinsèque, étant entendu que les dérivations par rapport aux cordonnées de l'espace tangent et cotangent sont toujours écrites par rapport à des bases duales de ces deux espaces.

Par ailleurs, on voit facilement qu'il suffit que l'inegalité souhaitée soit remplie localement pour un certain système de çordonnées pour qu'elle le soit pour tout système de cordonnées.

Définition (II. 2). Soit m un nombre réel. $\mathcal{L}(X;m)$ est l'espace des opérateurs linéaires continus A de $\mathcal{D}(X)$ dans $\mathcal{D}(X)$, qui se prolongent continûment de $\mathcal{D}(X)$ dans $\mathcal{D}(X)$, et pour tout s réel, de $H^s_{comp}(X)$ dans $H^{s-m}_{comp}(X)$, et qui, en outre, sont très réguliers, c'est à dire vérifient la condition suivante: si T est une distribution sur X de classe C^∞ dans un ouvert Ω de X, alors la distribution AT est de classe C^∞ dans Ω.

Remarques:

Tous les espaces de distributions considérés ici sont de sous-espaces de $\mathcal{D}'(X)$, espace des courants pairs de degré 0, $\mathcal{D}(X)$ étant l'espace des fonctions (à valeurs complexes) de classe C^∞ sur X et à support compact, $\mathcal{E}(X)$ l'espace des fontions de classe C^∞ et $\mathcal{E}'(X)$ l'espace de courants à support compact.

Il résulte facilement de la définition que si $A \in \mathcal{L}(X;m)$, alors A opère de $\mathcal{E}'(X)$ dans $\mathcal{E}'(X)$, de $H^s_{loc}(X)$ dans $H^{s-m}_{loc}(X)$ et de $\mathcal{E}(X)$ dans $\mathcal{E}(X)$.

Il convient de remarquer, pour un usage ultérieur, que dans la définition (II. 2), on peut supprimer l'hypothèse que A opère de $\mathcal{D}'(X)$ dans $\mathcal{D}'(X)$ (la dernière condition étant alors remplacée par la condition: $T \in \mathcal{E}'(X)$ et T est C^∞ dans Ω entraînent AT est C^∞ dans Ω) si pour tout compact K de X on peut trouver un compact L de X tel que $T \in \mathcal{D}(X)$ et $T=0$ sur L entraînent $AT = 0$ sur K.

En effet, moyennat cette condition, AT s'annulera sur $\mathring{K}$

si $T \in \mathcal{E}'(X)$ et s'annule dans un voisinage de L , puisque la régularisation n'augmente pas trop le support; cela étant, soit (φ_i) une partition C^ω de l'unité sur X, dénombrable et localement finie; alors si $T \in \mathcal{E}'(X)$ la famille $(A(\varphi_i T))$ est une famille de distribution localement finie (le support de $A(\varphi_i T)$ ne pouvant couper un compact K si le support de φ_i ne coupe pas le compact L associé) et en définissant $AT = \sum_i A(\varphi_i T)$ on obtient le prolongement voulu de A à $\mathcal{D}'(X)$; l'inclusion supp. sing. $AT \subset$ supp. sing. T pour toute $T \in \mathcal{D}'(X)$ est évidente.

Enfin, signalons que $\mathcal{L}(X;-\infty) = \bigcap_m \mathcal{L}(X;m)$ n'est que l'espace des opérateurs linéaires continus de $\mathcal{E}'(X)$ dans $\mathcal{D}(X)$ et de $\mathcal{D}'(X)$ dans $\mathcal{E}(X)$; nous désignerons ces opérateurs sous le nom d'opérateurs régularisants.

Proposition (II. 1). $C^\omega(T^*(X);m)$ est complet pour sa structure de groupe topologique (non séparé) pour laquelle un système fondamental de voisinages de 0 est constitué par les sous-espaces $C^\omega(T^*(X);k)$ (k réel $\leq$ m).

Cette proposition, et la suivante, permettent de sommer des séries de symboles ou d'opérateurs dont les ordres tendent vers $-\infty$. Elles ont été données en même temps, la première par Hörmander ([1]), et la deuxième par nous-mêmes ([2]).

Proposition (II. 2). $\mathcal{L}(X;m)$ est complet pour sa structure de groupe topologique pour laquelle un système fondamental de voisinages de 0 est constitué par les sous-espaces $\mathcal{L}(X;k)$ (k réel $\leq$ m).

La démonstration repose ici sur le théorème de Mittag-Leffler (cf. Bourbaki, Topologie générale, chapitre 2), les lemmes

(II. 1) et (II. 2) sont des intermédiaire à cette démonstration.

Nous commençons par définir, pour tout compact K de X et tout nombre réel m, l'espace $\mathcal{M}_K(X;m)$ suivant: c'est l'espace des opérateurs A de $\mathcal{E}(X)$ dans $\mathcal{E}'(X)$ qui vérifient les deux conditions suivantes:

(1) $\forall T \in \mathcal{E}(X)$, supp $(T) \subset \complement K \Longrightarrow AT = 0$

(2) $\forall s, t \in \mathbb{R}$, $t > m$, A opère de $H^s_{loc}(X)$ dans $H^{s-t}_{loc}(X)$.

Un tel opérateur A opère évidemment de $\mathcal{E}(X)$ dans $\mathcal{E}(X)$ et de $\mathcal{E}'(X)$ dans $\mathcal{E}'(X)$. On peut mettre sur $\mathcal{M}_K(X;m)$ une structure d'espace de Fréchet de la façon suivante: soit K' un compact de X, $K \subset \mathring{K}'$; quels que soient s et t, l'espace d'application linéaires continues $\mathcal{L}(H^s_{K'}(X), H^{s-t}_{loc}(X))$ est un espace de Fréchet pour la topologie de la convergence uniforme sur les parties bornées de $H^s_{K'}(X)$. On met sur $\mathcal{M}_K(X;m)$ la topologie borne supérieure des topologies induites par ces espaces, lorsque s et t parcourent $\mathbb{Q}$, avec $t > m$.

<u>Lemme (II. 1)</u>. Si $t < m$, $\mathcal{M}_K(X;t)$ est dense dans $\mathcal{M}_K(X;m)$.

Il suffit de construire une suite $(\mathcal{R}_K)$ d'operateurs opérant pour tout s de $H^s_{loc}(X)$ dans $H^{s+m-t}_{loc}(X)$ et convergeant simplement vers l'opérateur identique de $H^s_{loc}(X)$, comme on peut le voir par application du théorème de Banach-Steinhaus et du lemme de Rellich.

Par choix d'un système fini de cartes, on se ramène alors au problème suivant: soient L_1 et L_2 deux compacts de $\mathbb{R}^n$, $L_1 \subset \mathring{L}_2$; montrer qu'il existe une suite d'opérateurs sur $\mathbb{R}^n$ convergeant simplement pour tout s vers l'application identique de

$H^s_{L_1}(\mathbb{R}^n)$ dans $H^s_{L_2}(\mathbb{R}^n)$ et opérant de $H^s_{L_1}(\mathbb{R}^n)$ dans $\mathcal{E}_{L_2}(\mathbb{R}^n)$. Or si (γ_k) est la suite standard de fonctions de $\mathcal{E}(\mathbb{R}^n)$ convergeant vers la distribution de Dirac δ, la suite des opérateurs de convolution associés répond à la question.

<u>Lemme (II. 2).</u> Soient K et K' deux compacts de $X, K \subset \overset{\circ}{K'}, (m_j)_{j>0}$ une suite strictement décroissante de nombres réels tendant vers $-\infty$, et pour tout j, un opérateur $A_j \in \mathcal{L}(X;m_j)$ à bisupport compact dans K, c'est-à-dire tel que

$$\forall T \in \mathcal{E}(X),\ \text{supp.}(A_jT) \subset K,\ \text{et:}$$
$$\forall T \in \mathcal{D}(X),\ \text{supp.}(T) \subset \complement K \Longrightarrow A_jT = 0.$$

Il existe alors $A \in \mathcal{L}(X;m_0)$, à bisupport compact dans K', tel que, quel que soit k, $A - \sum_{j\leq k-1} A_j$ apppartienne à $\mathcal{L}(X;m_k)$.

En effet, soit pour tout, $W_k = \sum_{j\leq k-1} A_j + \mathcal{L}_K(X;m_k)$, que l'on munit de la métrique de $\mathcal{L}_K(X;m_k)$ transportée par translation D'après le lemme (II. 1) et le théorème de Mittag-Leffler, il existe $B \in \bigcap_k W_k$ et si $\varphi \in \mathcal{D}_{K'}(X)$ est égale à 1 dans un voisinage de K, $A=(\varphi)B$ répond à la question ((φ) est ici l'opérateur de multiplication par φ). On ramène à ce lemme le cas général par partition de l'unité.

J. Bokobza

CHAPITRE III

OPERATEURS PSEUDO-DIFFERENTIELS SUR UNE VARIETE DIFFERENTIABLE.

Soient v une linéarisation de X et Ω un domaine de v.

Quels que soient x et y apparténant à X, l'application linéaire

$$1_x^y : T_x^*(X) \longrightarrow T_y^*(X)$$

définit, en passant à la puissance extérieure n-ième, une application linéaire $\Lambda^n 1_x^y : \Lambda^n T_x^*(X) \longrightarrow \Lambda^n T_y^*(X)$, que l'on peut aussi interpréter comme un élément de $\Lambda^n T_x(X) \otimes \Lambda^n T_y(X)$, et une section $\Lambda^n 1$ de classe C du fibré ${}^n T(X)$ ${}^n T\,(X)$ au-dessus de XxX (il s'agit bien sûr ici du produit tensoriel externe).

Au-dessus de Ω, il y a également une section canonique (relativement à v) du fibré $\Lambda_t^n T(X) \otimes \Lambda_t^n T^*(X)$, où $\Lambda_t^n T(X)$ (resp. $\Lambda_t^n T^*(X)$) est l'espace des n-vecteurs tordus (resp. n-covecteurs tordus), section qui se déduit de la précedente puisqu'une orientation de Ω_x est définie canoniquement par la donnée d'une orientation de X au point x (en effet $y \longrightarrow v(x,y)$ définit un difféomorphisme de Ω_x sur un ouvert de $T_x(X)$).

On désignera la valeur de cette dernière section au point (x,y) par la notation $d\xi \otimes \text{dét} \dfrac{\partial v(x,y)}{\partial y}\, dy$. C'est du reste son expression en coordonnées locales, lorsque des coordonnées locales sont choisies dans des voisinages de x et de y de façon compatible quant à l'orientation définie par la carte $\Omega_x \longrightarrow T_x(X)$ et par n'importe laquelle des deux orientations possibles de $T_x(X)$.

Si f est une fonction localement sommable sur X, à support dans Ω_x, on désignera par $d\xi \int_{\Omega_x} f(y) \det \frac{\partial v(x,y)}{\partial y} dy$ l'intégrale sur Ω_x de $f(y) d\xi \wedge \det \frac{\partial v(x,y)}{\partial y} dy$, qui est une forme impaire de degré n sur Ω_x, à valeurs dans l'espace des n-vecteurs tordus en x: cette intégrale est un n-vecteur tordu en x, qui peut donc servir de mesure sur $T_x^*(X)$.

Définition (III. 1). Soient Ω un domaine d'une linéarisation v de X, et α une fonction C^∞ sur XxX, à support contenu dans Ω, et égale à 1 dans un voisinage de la diagonale Δ de XxX.

Si $F \in C^\infty(T^*(X);m)$, m étant un nombre réel quelconque, $A = \mathcal{O}_{v,\alpha}[F]$ est l'opérateur défini sur $\mathcal{E}(X)$ par

$$(A\varphi)(x) = \int_{T_x^*(X)} F(x,\xi) d\xi \int_{\Omega_x} \alpha(x,y)\varphi(y) e^{-2i\pi\langle v(x,y),\xi\rangle} \det \frac{\partial v(x,y)}{\partial y} dy.$$

On notera par $(\mathcal{F}\varphi)(x,\xi)$ la valeur en (x,ξ) de la seconde intégrale. D'autre part on notera v_x le difféomorphisme de Ω_x sur un ouvert de $T_x(X)$ défini par $v_x(y) = v(x,y)$ et w_x le difféomorphisme réciproque; $\tilde{\alpha}(x,\eta)$ sera la fonction définie sur T(X) par $\tilde{\alpha}(x,\eta) = \alpha(x, w_x(\eta))$ si $(x,\eta) \in v(\Omega)$ et nulle en dehors de $v(\Omega)$. On peut alors écrire aussi

$$(\mathcal{F}\varphi)(x,\xi) = d\xi \int_{T_x(X)} \tilde{\alpha}(x,\eta)\,\varphi(w_x(\eta)) e^{-2i\pi\langle\eta,\xi\rangle} d\eta.$$

Proposition (III. 1). Avec les notations de la définition (III. 1), A opère de $\mathcal{E}(X)$ dans $\mathcal{E}(X)$ et de $\mathcal{E}'(X)$ dans $\mathcal{E}'(X)$.

On va montrer que si s et t sont deux entiers tels que $s-m-n > t \geq 0$, $s > 0$, A opère de $H^s_{comp}(X)$ dans $C^t_{comp}(X)$, espace des fonctions de classe C^t à support compact, ce qui montrera la proposition.

Définissons une structure riemannienne sur X par le choix d'un produit scalaire $(.|.)$ sur $T^*(X)$, prolongé de façon $\mathbb{C}$-bilinéaire sur le complexifié de $T^*(X)$. On a, pour $\xi \neq 0$:

$$e^{-2i\pi\langle v(x,y),\xi\rangle} = -\frac{1}{2i\pi|\xi|^2}\left(\xi \mid (l^y_x)^{-1} d_y e^{-2i\pi\langle v(x,y),\xi\rangle}\right),$$

d'où après une intégration par parties (effectué sur l'espace tangent):

$$(\tilde{\mathcal{F}}\varphi)(x,\xi) = \frac{1}{2i\pi}\, d\xi \int_{\Omega_x} e^{-2i\pi\langle v(x,y),\xi\rangle} \times$$

$$\times \left(\xi \mid (l^y_x)^{-1} d_y\left(|\xi|^{-2}\alpha(x,y)\varphi(y)\right)\right) \det \frac{\partial v(x,y)}{\partial y}\, dy,$$

ce qui donne, on itérant s fois le procédé,

$|(\tilde{\mathcal{F}}\varphi)(x,\xi)| \leq C\,|\xi|^{-s}\,\|\varphi\|_{H^s}$ pour $\xi \neq 0$, et comme on a

$|(\tilde{\mathcal{F}}\varphi)(x,\xi)| \leq C\,\|\varphi\|_{H^0}$ pour $|\xi| \leq 1$,

$$|(A\varphi)(x)| \leq C\,\|\varphi\|_{H^0} + C\int_{|\xi|>1} |F(x,\xi)|\,|\xi|^{-s}\,\|\varphi\|_{H^s}\, d\xi \leq C_1\|\varphi\|_{H^s}$$

puisque $s-m-n > 0$; enfin, en appliquant un opérateur différentiel P

d'ordre $\leq$ t à $A\varphi$ écrite sous sa forme initiale, on voit que l'on a encore

$$|(PA\varphi)(x)| \leq C \|\varphi\|_{H^s}$$

Proposition (III. 2). Soit l'opérateur défini dans la définition (III. 1), et soit $h: Y \longrightarrow X$ un difféomorphisme. Si $h^* A$ est l'opérateur défini sur $\mathcal{D}(Y)$ par $(h^* A)(\varphi) = A(\varphi o h^{-1}) o h$, on a

$$h^* A = \Theta_{h^* v, h^* \alpha}[h^* F] \ ,$$

avec

$$(h^* \alpha)(x, y) = \alpha(hx, hy), \quad (h^* v)(x, y) = dh(x)^{-1} \; v(hx, hy) \quad \text{et}$$

$$(h^* F)(x, \xi) = F(hx, {}^t dh(x)^{-1} \xi).$$

Il suffit de faire un changement de variables dans les intégrales qui définissent $h^* A$.

Définition (III. 2). Soit A un opérateur linéaire continu de $\mathcal{D}(X)$ dans $\mathcal{E}(X)$. Soient v une linéarisation de X et α une fonction C^∞ sur XxX, à support dans un domaine de v , égale à 1 dans un voisinage de Δ . Le symbole de A par rapport à (v, α) est la fonction (de classe C^∞) sur $T^*(X)$ définie par

$$\sigma_{v, \alpha}(A)(x, \xi) = A_y(\alpha(x, y) e^{2i\pi \langle v(x, y), \xi \rangle})(x),$$

la notation A_y signifiant que l'opérateur A agit sur la

fonction qui suit considérée comme fonction de y , les autres variables étant fixées.

Proposition (III. 3). Soient $F \in C^{\infty}(T^*(X);m)$, v et v_1 deux linéarisations de X, α et α_1 deux fonction C^{∞} sur XxX à support dans des domaines respectifs de v et v_1, égales à 1 dans un voisinage de Δ , et $A = \Theta_{v,\alpha}[F]$. Alors $\sigma_{v1,\alpha 1}(A)$ appartient à $C^{\infty}(T^*(X);m)$ et ne dépend pas de α ni de α_1 à un élément de $C^{\infty}(T^*(X);-\infty)$ près. En fait, si ψ_x est la fonction définie sur $v_x(\Omega) \subset T_x(X)$ qui fait passer de $v(x,y)$ à $v_1(x,y)$, on a:

$$\sigma_{v_1,\alpha_1}(A)(x,\xi) \sim \sum_{|r| \geq 0} \frac{1}{r!} \partial_\xi^r F(x,\xi)\, D_\eta^r \left\{ e^{2i\pi < \psi_x(\eta)-\eta\,,\,\xi >} \right\}(\eta=0),$$

et en particulier $\sigma_{v,\alpha}(A) \sim F$.

Avant de démontrer cette proposition, remarquons que la somme

$$\sum_{|r|=k} \frac{1}{r!} \partial_\xi^r F(x,\xi) D_\eta^r \left\{ e^{2i\pi < \psi_x(\eta)-\eta,\xi >} \right\} (\eta=0$$

est bien définie indépendamment du système de cordonnées choisi et appartient a $C^{\infty}(T^*(X); m-\left[\frac{k+1}{2}\right])$, $\left[\frac{k+1}{2}\right]$ désignant la partie entière de $\frac{k+1}{2}$; en effet $D_\eta^r \left\{ e^{2i\pi < \psi_x(\eta)-\eta,\xi >} (\eta=0)\right.$ est un polynôme en ξ de degré $\leq \left[\frac{|r|}{2}\right]$ dont les coefficients sont des fonctions C^{∞} de x , comme il résulte du fait que $\psi_x(\eta) - \eta$ s'annule ainsi que ses dérivées du premier ordre en $\eta = 0$.

La série est alors convergente au sens de la proposition (II. 1).

Posons alors $\alpha_2(x,y) = \alpha(x,y)\,\alpha_1(x,y)$; d'autre part choisissons une structure riemannienne sur X, et une fonction de classe C^∞ sur T(X), égale à 1 pour $|\eta| \leq \frac{1}{2}$ et à 0 pour $|\eta| \geq 1$ et posons

$$\gamma(x,y,\xi) = \beta(x, |\xi|^{1/4} v(x,y)).$$

On écrit alors $\sigma_{v_1, \alpha_1}(A) = G_1 + G_2$, avec

$$G_1(x,\xi) = \int_{T_x^*(X)} F(x,\zeta)\, d\zeta \int \alpha_2(x,y)\,\gamma(x,y,\xi) \times$$
$$\times\, e^{-2i\pi <v(x,y),\zeta> + 2i\pi <v_1(x,y),\xi>}\, \det \frac{\partial v(x,y)}{\partial y}\, dy$$

G_2 s'obtenant en remplaçant dans cette expression γ par $1-\gamma$.

On démontre alors que $G_2 \in C^\infty(T^*(X); -\infty)$ en écrivant que

$$1 - \beta(x,\eta) = |\eta|^{4N}\, \chi(x,\eta)$$

N étant un entier arbitrairement grand et en intégrant par parties, au moyen de la carte définie par le difféomorphisme $(v_1)_x$, la deuxième intégrale.

Quant à G_1, il s'écrit, après les changements de variables

$$\zeta \longmapsto \xi + \zeta \quad , \quad \eta = v(x,y)\,,$$

$$G_1(x,\xi) = \sum_{|\alpha| \leq 2k-1} \frac{1}{\alpha!}\, \partial_\xi^\alpha F(x,\xi)\, D_\eta^\alpha \{ e^{2i\pi <\psi_x(\eta) - \eta, \xi>} \}(\eta = 0) +$$
$$+\, R_k(x,\xi)$$

où $R_k(x, \xi) =$

$$\iint_{T_x^*(X)} \Big[F(x, \xi+\zeta) - \sum_{|t|\leq 2k-1} \frac{1}{t!} \partial_\xi^t F(x,\xi)\,\zeta^t \Big] d\zeta \times$$

$$\times \int_{T_x(X)} \tilde{\alpha}_2(x,\eta)\,\beta(x, |\xi|^{1/4}\eta)\, e^{-2i\pi\langle\eta,\zeta\rangle + 2i\pi\langle \psi_x(\eta)-\eta, \xi\rangle} d\eta .$$

Il s'agit de prouver que quels que soient p et q, $D_x^p \partial_\xi^q R_k(x, \xi)$ est majoré, x restant dans un compact, par telle puissance de ξ que l'on veut, pourvu que k soit assez grand.

Mais d'après la formule de Leibniz, une dérivée de R_k est somme d'expressions analogues à celle de R_k, mais où F est remplacée par une dérivée de F et $\tilde{\alpha}_2(x,\eta)$ par un polynôme en ξ et $|\xi|^{1/4}$ dont les coefficients sont des fonctions C^∞ de (x, η). Il suffit donc d'obtenir une estimation de R_k.

On sèpare R_k en deux: $R_k^1(x, \xi) = \int_{|\zeta| < \frac{|\xi|}{2}}$ et $R_k^2(x, \xi) = \int_{|\zeta| > \frac{|\xi|}{2}}$.

Pour simplifier les notations posons

$$\varphi(x,\xi,\eta) = \tilde{\alpha}_2(x,\eta)\,\beta(x,|\xi|^{1/4}\eta)\, e^{2i\pi\langle \psi_x(\eta)-\eta, \xi\rangle}$$

$\hat{\varphi}(x, \xi, \zeta)$ désignant sa transformée de Fourier par rapport à η, calculée au point ζ. Dans le support de $\beta(x, |\xi|^{1/4}\eta)$, on a $\eta \leq |\xi|^{-1/4}$, et $\psi_x(\eta) - \eta$ s'annulant ainsi que ses dérivées premières en $\eta = 0$, une dérivée d'ordre j (par rapport à η) de $\varphi(x, \xi, \eta)$ est majoré par $C|\xi|^{3j/4}$, où C est une constante indépendante de ξ et de x lorsque ce dernier parcourt un compact. Ceci prouve que, quel que soit ν entier positif,

$$|\hat{\varphi}(x,\xi,\zeta)| \leq C|\zeta|^{-2\nu}|\xi|^{3\nu/2}$$

pour $\zeta \neq 0$.

Pour $|\zeta| > \frac{|\xi|}{2}$, on a

$$\left|F(x,\xi+\zeta) - \sum_{|t|\leq 2k-1} \frac{1}{t!}\partial_\xi^t F(x,\xi)\zeta^t\right| \leq C|\zeta|^{2k+\sup(m,0)}$$

et $|\zeta|^{-2\nu} \leq C|\zeta|^{-\nu/4}|\xi|^{-7\nu/4}$, d'où

$$R_k^2(x,\xi) \leq C|\zeta|^{-\nu/4}\int_{|\zeta|>\frac{|\xi|}{2}} |\zeta|^{2k+\sup(m,0)-\nu/4} d\xi .$$

Pour évaluer R_k^1, on écrit pour $|\zeta| < \frac{|\xi|}{2}$:

$$\left|F(x,\xi+\zeta) - \sum_{|t|\leq 2k-1} \frac{1}{t!}\partial_\xi^t F(x,\xi)\zeta^t\right| \leq$$

$$\leq 2\sum_{|t|=2k} \frac{1}{t!}|\partial_\xi^t F(x,\xi+\theta\zeta)||\zeta|^{2k} \leq C|\xi|^{m-2k}|\zeta|^{2k}$$

d'où

$$|R_k^1(x,\xi)| \leq C|\xi|^{m-2k}\int_{|\zeta|<\frac{|\xi|}{2}} |\zeta|^{2k}|\hat{\varphi}(x,\xi,\zeta)|\,d\zeta \leq$$

$$\leq C|\xi|^{3k/4+n/2}\|\varphi(x,\xi,\cdot)\|_{H^k}|\xi|^{m-2k}$$

et finalement $|R_k^1(x,\xi)| \leq C|\xi|^{m+\frac{n}{2}-\frac{k}{4}}$.

Proposition (III. 4) . Soient A un opérateur régularisant (c'est-à-dire opérant de $\mathcal{E}'(X)$ dans $\mathcal{D}(X)$ et de $\mathcal{D}'(X)$ dans $\mathcal{E}(X)$), v une linearisatio de X et α une fonction C^∞ sur XxX, à support dans un domaine de v, égale à 1 dans un voisinage de Δ. Alors $\sigma_{v,\alpha}(A) \in C^\infty(T^*(X);-$

Les estimations qu'il faut prouver étant de nature locale en x, étant donné que, lorsque x reste dans un compact, $\alpha(x,y)$ ne peut être non nul que si y reste dans un compact, on peut, après partition finie de l'unité, supposer que $X = \mathbb{R}^n$. Remarquons ainsi que, puisque

$$\partial_\xi^{(j)} \tilde{\sigma}_{v,\alpha}(A)(x,\xi) = 2i\pi A_y \left\{ \alpha(x,y) v_j(x,y) e^{2i\pi\langle v(x,y),\xi\rangle} \right\}(x) \quad \text{et que}$$

$$D_x^{(j)} \tilde{\sigma}_{v,\alpha}(A)(x,\xi) = \left[(D^{(j)}A)_y \left\{ \alpha(x,y)\, e^{2i\pi\langle v(x,y),\xi\rangle} \right\}(x) + \right.$$

$$+ A_y \left\{ D_x^{(j)} \alpha(x,y) e^{2i\pi\langle v(x,y),\xi\rangle} \right\}(x)$$

$$+ \sum_k \xi_k A_y \left\{ \alpha(x,y) \frac{\partial v_k(x,y)}{\partial x_j} e^{2i\pi\langle v(x,y),\xi\rangle} \right\}(x),$$

on peut se borner à montrer que, quel que soit l'entier N, $|\xi|^N \tilde{\sigma}_{v,\alpha}(A)(x,\xi)$ reste borné indépendamment de ξ lorsque x parcourt un compact. Puisque A est régularisant, il suffit de montrer que, dans les conditions données $|\xi|^N \alpha(x,y) e^{2i\pi\langle v(x,y),\xi\rangle}$ reste borné dans $\mathcal{E}'_y(X)$, autrement dit que si φ reste dans une partie bornée $\mathcal{E}(X)$, l'intégrale

$$|\xi|^N \int \alpha(x,y) e^{2i\pi\langle v(x,y),\xi\rangle} \varphi(y) dy$$

reste borné. En remarquant que

$$\sum_j (1_x^y(\xi))_j\, D_y^{(j)} \left(e^{2i\pi\langle v(x,y),\xi\rangle}\right) = e^{2i\pi\langle v(x,y),\xi\rangle} |1_x^y(\xi)|^2,$$

on peut écrire cette intégrale, après intégration par parties, sous la forme

$$|\xi|^N \int e^{2i\pi\langle v(x,y),\xi\rangle} \sum_j D_y^{(j)} \left(\frac{\alpha(x,y)\, \varphi(y) (1_x^y(\xi))_j}{|1_x^y(\xi)|^2} \right) dy,$$

et il suffit de répéter l'opération N fois pour obtenir le résultat.

Proposition (III. 5). Soit A un opérateur linéaire continu de $\mathcal{D}(X)$ dans $\mathcal{E}(X)$, et soient v et α une linéarisation de X et une fonction C^∞ sur $X \times X$, à support dans un domaine de v, égale à 1 dans un voisinage de Δ. Alors quelle que soit $\varphi \in \mathcal{D}(X)$:

$$A_z\left[(\alpha(x,z))^2\varphi(z)\right](x) = \int_{T^*_x(X)} \sigma_{v,\alpha}(A)(x,\xi)\,d\xi_x$$

$$\int_{\Omega_x} \alpha(x,y)\,\varphi(y)e^{-2i\pi\langle v(x,y),\xi\rangle}\,\text{dét}\,\frac{\partial v(x,y)}{\partial y}\,dy.$$

Remarquons tout de suite que si A se prolonge de $\mathcal{E}'(X)$ dans $\mathcal{D}'(X)$, de $\mathcal{S}'(X)$ dans $\mathcal{S}'(X)$, de $\mathcal{S}(X)$ dans $\mathcal{E}(X)$ et est très régulier alors l'opérateur B défini par

$$(B\varphi)(x) = A_z\left[(\alpha(x,y))^2\varphi(z)\right](x)$$

diffère de A par un opérateur régularisant.

La proposition se démontre facilement en remarquant que la transformée de Fourier de la fonction $\eta \mapsto \alpha(x, w_x(\eta))\varphi(w_x(\eta))$, où $w_x = v_x^{-1}$, est égale à

$$d\xi\int_{\Omega_x} \alpha(x,y)\,\varphi(y)e^{-2i\pi\langle v(x,y),\xi\rangle}\,\text{dét}\,\frac{\partial v(x,y)}{\partial y}\,dy,$$

et que $(\alpha(x,z))^2\varphi(z)$ peut donc s'écrire sous la forme

$$\alpha(x,z)e^{2i\pi\langle v(x,z),\xi\rangle}\,d\xi\int \alpha(x,y)\,\varphi(y)e^{-2i\pi\langle v(x,y),\xi\rangle}\,\text{dét}\,\frac{\partial v(x,y)}{\partial y}\,dy.$$

<u>Proposition (III. 6).</u> Si $F \in C^{\infty}(T^*(X);m)$, alors $\Theta_{v,\alpha}[F] \in \mathcal{L}(X;m)$.

Il suffit de prouver qu'il en est ainsi pour un opérateur $A = (\psi_1)\Theta_{v,\alpha}[F](\psi_2)$, où la réunion des supports compacts de ψ_1 et ψ_2 admet un voisinage ouvert Y (non nécessairement connexe) domaine d'un système de coordonnées. L'opérateur A, qui est à bisupport compact dans Y, admet une restriction A_Y et il s'agit de prouver que $A_y \in \mathcal{L}(X;m)$.

On est donc ramené au cas d'un ouvert de $\mathbb{R}^n$ et l'opérateur obtenu est donné par une formule analogue à la formule habituelle (définition (III. 1) pour une linéarisation canonique v_o sur $\mathbb{R}^n$ définie par $v_o(x,y)=y-x$.

La proposition (III. 3) permet d'évaluer le symbole de cet opérateur par rapport à v_o et de voir qu'il a les propriétés requises pour être le symbole d'un opérateur pseudo-differentiel sur $\mathbb{R}^n$, opérant de H^s dans H^{s-m}; on termine la demonstration grâce à la proposition (III. 5).

On deduit immédiatement des propositions précédentes que v et α étant fixées, l'espace vectoriel engendré par les opérateurs régularisants et par les opérateurs de la forme $\Theta_{v,\alpha}[F]$ où $F \in C^{\infty}(T^*(X);m)$ ne dépend pas de v ni de α : on définit ainsi une classe d'opérateurs pseudo-différentiels sur X que l'on notera PSD(X;m). Toujours d'après ce qui précède, il est évident que $PSD(X;m) \subset \mathcal{L}(X;m)$ et que v et α étant donnés, $\Theta_{v,\alpha}$ et $\sigma_{v,\alpha}$ définissent par passage aux quotients deux isomorphismes réciproques l'un de l'autre entre

$$C^{\infty}(T^{*}(X);m) \big/ C^{\infty}(T^{*}(X);-\infty)$$

et $\quad PSD(X;m) \big/ \mathcal{L}(X;-\infty)$

et ces isomorphismes ne dépendent pas de α .

D'autre part, v et α étant donnés, $\Theta_{v,\alpha}$ et $\sigma_{v,\alpha}$ définissent deux isomorphismes réciproques l'un de l'autre entre l'espace des polynômes $p(x,\xi)$ appartenant à $C^{\infty}(T^{*}(X);m)$ et l'espace des opérateurs différentiels d'ordre $\leq m$ sur X, à coefficients C^{∞}, et ces isomorphismes ne dépendent pas de α .

Proposition (III. 7). Soient G un groupe de Lie compact opérant sur X et $d\mu$ une mesure de Haar sur G de masse totale égale à 1. Soient v_1 une linéarisation de X et α_1 une fonction C^{∞} à support compact dans un domaine de v_1, égale à 1 dans un voisinage de Δ et soient $F \in C^{\infty}(T^{*}(X);m)$ et A un opérateur de $\mathcal{D}(X)$ dans $\mathcal{E}(X)$. Alors:

$v(x,y) = \int_G dg(g^{-1}x)v_1(g^{-1}x, g^{-1}y)d\mu(g)$ est une linéarisation G-invariante sur X et $\alpha(x,y) = \int_G \alpha_1(g^{-1}x, g^{-1}y)d\mu(g)$ est une fonction C^{∞} à support compact dans un domaine de v, égale à 1 dans un voisinage de Δ et G-invariante. En outre, si g G

$$\Theta_{v,\alpha}\left[g^{*}F\right] = g^{*}\,\Theta_{v,\alpha}[F]$$

et $\quad \sigma_{v,\alpha}(g^{*}A) = g^{*}\,\sigma_{v,\alpha}(A),$

où

$$(g^{*}F)(x,\xi) = F(gx, {}^{t}dg(x)^{-1}\xi)$$

$$\text{et} \quad (g^{*}A)(\varphi)(x) = A(\varphi \circ g^{-1}) \circ g.$$

La démonstration est purement formelle.

On en déduit qu'un opérateur pseudo-differentiel sur X est G-invariant à un opérateur régularisant près si et seulement si son symbole relativement à un v et un α G-invariants est G-invariant à un élément de $C^{\infty}(T^{*}(X);-\infty)$près.

Toute cette description s'étend aux opérateur des sections d'un fibré vectoriel E^1(complexe) dans les sections d'un autre fibré vectoriel E^2, moyennant les modifications suivantes:

a). La fonction $F(x,\xi)$ de la définition (III. 1) doit être remplacée (comme il est bien connu du reste) par une fonction à valeurs dans $\mathcal{L}(E^1;E^2)$.

b). Pour donner un sens aux formules des définitions (III. 1) et (III. 2), il convient d'introduire un "transport local" du fibré E^1 : par celà, nous entendons une section C^{∞} du produit tensoriel externe $E^{1*} \boxtimes E^1$ coincidant avec l'identité sur la diagonale de XxX, l'existence d'un tel transport local étant assurée sans aucune condition particulière relative à E^1.

c). La proposition (III. 7) s'étend à condition de supposer que G opère également sur E^1 et E^2 de façon compatible avec son opération sur X et que le transport local de E^1 est G-invariant.

J. Bokobza

Bibliographie:

1 Hörmander, L., Pseudo-differential operators, Comm. Pure Appl. Math. 18 (1965) pp. 501-517.

2 Unterberger, A., et Bokobza, J., Les opérateurs de Calderón-Zygmund précisés, C. R. Acad. Sc. Paris, tome 259, 1964, pp. 1612-1614, tome 260, 1965, pp. 3265-3267.

CENTRO INTERNAZIONALE MATEMATICO ESTIVO

(C. I. M. E.)

L. BOUTET DE MONVEL

PSEUDO-DIFFERENTIAL OPERATORS AND ANALYTIC FUNCTIONS

Corso tenuto a Stresa dal 26 Agosto al 3 Settembre 1968

CHAPITRE I OPERATEURS PSEUDO DIFFERENTIELS ANALYTIQUES

by. L. Boutet de Monvel (Universitè d'Alger)

§ 0- Introduction Le but de ce chapitre est de décrire une classe d'opérateurs pseudo-différentiels, sur les variétés analytiques réelles, qui se comporte bien vis à vis des fonctions analytiques. En gros ces opérateurs ont les propriétés suivantes

1. Un opérat r pseudo-différentiel analytique est est en particulier un opérateur pseudo-différentiel ordinaire, et le calcul symbolique (de Calderon, Kohn, Nirenberg) marche encore.

2. Si deux opérateurs pseudo-différentiels analytiques ont même symbole, ils diffèrent par un opérateur à noyau analytique (de la forme $f \to \int k(x,y)\ f(y)\ dy$, où $k(x,y)\ dy$ est analytique en x et y)

3. Un opérateur pseudo-différenriel analytique conserve l'analyticité localement : si P est un opérateur pseudo-différentiel analytique, si f est une distribution à support compact, analytique au voisinage d'un point x, $P(f)$ est aussi analytique au voisinage de x.

4. Composition : si V est une variété analyrique réelle compacte, P et Q deux opérateurs pseudo-différentiels analytiques sur V, $P \circ Q$ est encore un opérateur pseudo-différentiel analytique sur V.

Si V n'est pas compacte, il y a une petite difficulté du fait que les noyaux de P et Q sont analytiques hors de la diagonale de $V \times V$, de sorte qu'en général on ne peut pas composer parceque les intégrales qui interviennent n'ont aucune raison d'être convergentes à l'infini . Il y a quand même un résultat local , décrit au théorème 4

5. Si P est un opérateur pseudo-différentielanalytique elliptique , il possède une parametrix qui est aussi un opérateur pseudo-différentiel analytique .

Ces opérateurs permettent de prouver le résultat suivant :

Théorème 1.1 Soit P un opérateur pseudo-différentiel analytique elliptique (sur une variété analytique réelle V) ; soit f une distribution à support compact . Si P(f) est analytique au voisinage d'un point x , f elle-même est analytique au voisinage de x .

§ 1 - Définitions Dans ce paragraphe , nous décrivons localement le noyau de nos opérateurs : V est un ouvert de R^n . L'opérateur pseudo-différentiel P a pour noyau la distribution P(x, y) dy où pour x - y petit , P(x, y) dy admet le developpement asymptotique :

$$(1.1) \quad P(x,y) \sim \sum P_k(x, x-y)$$

où $P_k(x, z)$ est pseudo-homogéne (de degré $-n-d+k$) en z si deg P = d)

(Rappelons que la distribution T(z) est pseudo-homogéne de degré λ si pour λ non entier positif , c'est une distribution homogène de degré λ ; et pour λ entier positif on a

$$T(z) = Q(z) \,\mathrm{Log}\, |z| + f(z)$$

où f est homogène de degré λ , et Q est un polynôme homogène de degré λ).

Définition 1. On dit que P est un opérateur pseudo-différentiel analytique si

1. pour tout compact $K \subset V$, il existe un nombre $\varepsilon > 0$ tel que les $p_k(x, z)$ soient toutes holomorphes dans le domaine

$$d(x, K) < \varepsilon \quad (x \text{ complexe}), \quad |\operatorname{im} z| < \varepsilon\, |\operatorname{re} z|, \quad |z| < \varepsilon$$

et la série $\sum p_k(x, z)$ converge uniformément dans ce domaine.

2. le noyau-distribution $P(x, y)$ de P est une fonction analytique pour $x \neq y$, et diffère par une fonction analytique de la distribution $\sum p_k(x, z)$ au voisinage de $x = y$

(pour les termes p_k dont le degré $-n-d+k$ est entier plus petit que $-n$ - ce qui concerne un nombre fini de termes - on exige que p_k soit de la forme

$$p_k(x, z) = \sum a_\alpha(x)\, \delta^{(\alpha)}(z) + \mathrm{pf}_z\, f(x, z)$$

où les $a_\alpha(x)$ sont analytiques, $f(x, z)$ analytique dans le domaine ci-dessus, homogène de degré $-n-d+k$ en z, et a certains moments sphériques nuls ; $\delta^{(\alpha)}(z)$ désigne une dérivée d'ordre $d-k$ de la masse de Dirac)

§ 2- Premières propriétés

2.1. La définition 1.1 est invariante par changement de coordonnées analytique : la méthode exposée par R. T. Seeley pour les opérateurs pseudo-différentiels ordinaires marche ; c'est même encore plus simple dans notre cas, du fait qu'il s'agit de développement en séries convergentes

Ceci permet de définir les opérateurs pseudo-différentiels sur une variété analytique réelle : un opérateur pseudo-différentiel P sur une telle variété est analytique si son noyau-distribution P(x, y) dy est analytique hors de la diagonale x = y , et si localement , pour tout système de coordonées analytique , il est de la forme décrite dans la définition 1. 2

2. 2. Il est clair dans la définition 1. 2 que deux opérateurs analytiques qui ont même symbole diffèrent par un opérateur à noyau analytique

2. 3. On peut aussi donner une définition locale des symboles analytiques: un symbole d'opérateur pseudo-différentiel est dit analytique si pour tout point x de la variété de définition V , il existe un opérateur pseudo-différentiel analytique défini au voisinage de x , admettant ce symbole .

Avec cette définition , tout symbole analytique est (globalement) symbole d'un opérateur pseudo-différentiel analytique défini sur la variété toute entière : cela résulte du fait que la cohomologie , ou la cohomologie relative des fonctions analytiques réelles est nulle .

2. 4. Régularité D'abord un opérateur pseudo-différentiel analytique est évidemment un opérateur pseudo-différentiel ordinaire . En outre on a la propriété de régularité annoncée dans l'introduction , et même un peu mieux :

Théorème 1. 3 Si T est une fonctionelle analytique réelle à support compact , P (T) est bien définie ; c'est une hyper-fonction , analytique en dehors du support de T . Si de plus T est analytique au voisinage d'un point x , il en est de même de P (T) .

Voici une esquisse de la preuve : pour commencer , on met une topologie convenable sur l'espace des hyper-distributions à support dans un compact K , analytiques dans un ouvert U ; et sur l'espace des hyper-distributions analytiques dans U et V - K , de façon à en faire des espaces complets (cf. (1)) (il y a une légère difficulté provenant du fait que le support d'une hyper -

distribution n'en dépend pas ontinument ; mais il n'y a au une diffi ulté si on se restreint aux distributions ordinaires)

Le théorème se prouve alors ainsi

1) Il est vrai si P est l'opérateur de multiplication par une fonction analytique (et plus généralement si P est un opérateur différentiel à coefficients analytiques)

2) Il est vrai si P est un opérateur de convolution par une distribution T , o T est analytique en dehors de l'origine (c'est le cas quand la variété de définition de P est l'espace vectoriel R^n , et le noyau-distribution $P(x, y)\, dy$ de P ne dépend que de $x-y$) (cf. par exemple L. Schwartz , séminaire 1954-55 , exposé de Malgrange sur l'ellipticité analytique)

3) Il est evidemment vrai si le noyau-distribution de P est une fonction analytique .

Le cas général se déduit alors de 1) et 2) par un argument de produit tensoriel (quitte éventuellement à retrancher à P un opérateur à noyau-distribution analytique , le noyau-distribution de P est dans le produit tensoriel complété des espaces correspondant aux cas 1) et 2) , du moins localement).

1.5. Adjoint Si P est un opérateur pseudo-différentiel analytique, son adjoint l'est aussi .

Preuve : si P a pour noyau distribution $P(x, x-y)\, dy$, (où la variété de définition est R^n) , l'adjoint P^* de P a pour noyau-distribution $\overline{P}(y, y-x)$. Et pour $x-y$ petit , $\overline{P}(y, y-x)$ admet le développement en série :

$$\overline{P}(y, y-x) = \sum \left(\frac{\partial}{\partial x}\right)^{\alpha} \overline{P}(x, y-x) \quad (y-x)^{\alpha} / \alpha!$$

§3- Calcul symbolique ; représentation par intégrale de Fourier ; composés

Le théorème de composition annoncé au paragraphe O est le suivant :

Théorème 1.4 Soit V une variété analytique réelle ; P et Q deux opérateurs pseudo-différentiels analytiques sur V ; φ une fonction C^∞ à support compact, égale à 1 dans un ouvert $U \subset V$. Alors la restriction à U de $P\varphi Q$ est un opérateur pseudo-différentiel analytique.

La proposition est en réalité locale, et on la ramène immédiatement au cas où V est un ouvert de R^n. Il est alors commode de représenter P et Q au moyen d'une intégrale de Fourier, comme pour les opérateurs pseudo-différentiels ordinanres. C'est ce qui est décrit dans les trois propositions qui suivent.

On notera $p_k(x,\xi)$ la transformée de Fourier par rapport à z de la fonction $P_k(x,z)$ dans le développement asymptotique (1.1)'. Et on représente le symbole (complet) de P par la série formelle :

$$\sigma(P) = \sum p_k(x,\xi)$$

Proposition 1.5 $\sum p_k(x,\xi)$ est le symbole d'un opérateur pseudo-différentiel analytique si et seulement si pour tout compact K de V, il existe trois constantes ε, c, A telles que dans le domaine complexe

$$d(x,K) < \varepsilon \quad , \quad |\operatorname{im}\xi| < \varepsilon\, |\operatorname{re}\xi|$$

on ait l'inégalité

$$(1.2) \qquad |p_k(x,\xi)| \leq c\, A^k\, |\xi|^{d-k}\, k!$$

La démonstration est élémentaire le nombre ε de la proposition 1.5 est en gros le même que celui de la définition 1.1. La définition 1.1 implique que $P_k(x,z)$ croît aussi vite qu'une série géométrique pour $|z| \leq 1$ (dans le domaine complexe de la définition 1.1) :

$$|P_k(x,z)| \leq c\, A^k$$

le facteur $k!$ intervient tout naturellement quand on calcule la transformée de Fourier d'une fonction homogène de degré environ k : par exemple, si x^+ désigne la partie positive de x, la transformée de Fourier de la fonction $(x^+)^d$ est $d!\,(i\xi)^{-d-1}$.)

On a encore :

Proposition 1.6 Soit $\sum p_k(x,\xi)$ un symbole analytique sur un ouvert V de R^n. Soit $p(x,\xi)$ une fonction continue, analytique pour $\xi \neq 0$, et vérifiant la condition suivante :

Pour tout compact K de V, il existe trois constantes ε, c, A telles que p et les p_k soient analytiques dans le domaine complexe

$$d(x,K) < \varepsilon \quad , \quad |\mathrm{im}\,\xi| < \varepsilon\,|\mathrm{re}\,\xi|$$

et satisfont dans ce domaine à l'inégalité :

$$(1.3) \qquad \Big| p(x,\xi) - \sum_{k<N} p_k(x,\xi) \Big| \leq c\, A^N\, N!\, |\xi|^{d-N}$$

Alors l'opérateur pseudo-différentiel

$$f \longrightarrow (2\pi)^{-n} \int e^{ix.\xi}\, p(x,\xi)\, \hat{f}(\xi)\, d\xi$$

est analytique

La preuve consiste à majorer les dérivées successives de la fonction $P(x,z) - \sum P_k(x,z)$ pour z assez petit, et de vérifier ainsi qu'elle est analytique. (Comme ci-dessus, P -resp. P_k- est la transformée de Fourier inverse par rapport à ξ de p -resp p_k-) (cf. [1], prop. 2.5)

On a enfin

Proposition 1.7 Pour tout opérateur pseudo-différentiel analytique P sur un ouvert V de R^n, et pour tout ouvert U relativement compact dans V, il existe un opérateur à noyau-distribution analytique R sur U tel que l'opérateur $P' = P - R$ admette la représentation décrite dans la proposition 1.3

(Pour prouver cette proposition, on est amené à résoudre un problème de moments analogue à ceux qui interviennent quand on cherche à prolonger une fonction qui appartient à la classe de Gevrey G^2 ; cf. [1] , théorème 2.9 , et la bibliographie qui y est citée)

Voici pour terminer ce paragraphe une esquisse de la preuve du Théorème 1.4 :

(1) Le théorème est vrai si Q est la multiplication par une fonction analytique f (et plus généralement un opérateur différentiel à coefficients analytiques) :

si $P(x, x-y)$ est le noyau distribution de P , le noyau-distribution de $P \varphi Q$ est $P(x, x-y)\ f(y)$ dans U , et il suffit de développer f en série de Taylor pour $x-y$ petit :

$$f(y) = \sum f^{(\alpha)}(x)\ (y-x)^{\alpha} / \alpha!$$

(2) Le théorème est vrai si Q est un opérateur de convolution , et si P et Q admettent tous deux la représentation par intégrale de Fourier de la proposition 1.6 : alors $P(1-\varphi)Q$ est bien défini , et est un opérateur à noyau-distribution analytique dans U ; l'opérateur P Q admet la représentation :

$$P\,Q(f) = (2\pi)^{-n} \int e^{ix.\xi}\, p(x,\xi)\ q(\xi)\ \hat{f}(\xi)\ d\xi$$

et il est évident que la fonction $p(x,\xi)\ q(\xi)$ vérifie les conditions de la proposition 1.6

(3) Le théorème est évident quand P ou Q est un opérateur à noyau-distribution analytique

(4) Enfin , comme au paragraphe 2 , le cas général se déduit des trois cas ci-dessus par un argument de produits tensoriels topologiques.

§ 4- Parametrix - Norme formelle

La condition (1. 2) de la prop. 1. 5 , qui décrit les symboles analytiques , est équivalente à la suivante :

en posant $p_k^{\alpha\beta} = \sup_{|\xi|=1} \left| \left(\frac{\partial}{\partial x}\right)^{\alpha} \left(\frac{\partial}{\partial \xi}\right)^{\beta} p_k \right|$

la série

$$(1.4) \qquad N(p,T) = \sum c_k^{\alpha\beta} \; p_k^{\alpha\beta} \; T^{2k+|\alpha+\beta|}$$

est convergente

(plus exactement , son rayon de convergence , qui est une fonction de x , est bornée inférieurement loin de zéro , localement)

à condition que les coefficients c_k aient pour ordre de grandeur

$c_k \sim c^{k+|\alpha+\beta|} / k!\, \alpha!\, \beta!$

Si on choisit précisément

$$(1.5) \qquad c_k = 2\ (2n)^{-k}\ k! \ /\ (k+|\alpha|)!\ (k+|\beta|)!$$

on a en outre le résultat suivant :

Proposition 1. 8 Si la norme formelle $N(p,T)$ et les coefficients $c_k^{\alpha\beta}$ sont ceux des formules (1. 4) , (1. 5) , on a l'inégalité

$$N(p\circ q, T) \ll N(p,T)\ N(q,T)$$

où $p\circ q$ désigne le produit de p et q au sens de la composition et le signe $\ll$ signifie que le coefficient de T^j dans le deuxième membre majore celui de T^j dans le premier .

Corollaire 1. 9 Si $p = \sum p_k(x,\xi)$ est un symbole analytique elliptique (ie. p est inversible , ou , ce qui revient au même , le symbole principal p_0 est inversible) , alors le symbole inverse de p est analytique .

Preuve : soit q le symbole d'opérateur pseudo-différentiel

$$q = \sum q_k(x,\xi)$$

où tous les q_k sont nuls , sauf le premier qui est égal à p_0^{-1}

Il est clair que q est un symbole analytique ; et le composé des symboles p et q est de la forme

$$q \circ p = 1 - h$$

uù h est un symbole analytique de degré -1 .

Le problème est donc de prouver que le symbole inverse de $1 - h$ est analytique .

Or on a $(1 - h)^{-1} = \sum h^k$

et par suite , la norme formelle de $(1 - h)^{-1}$ est dominée par

$$N((1-h)^{-1}, T) \ll \sum N(h, T)^k = 1 / 1 - N(h, T)$$

c'est une série convergenteparceque la norme formelle de h est une série convergente , divisible par T .

Dans le cas des systèmes d'opérateurs pseudo-différentiels , on construit par le même procédé une parametrix à gauche ou à droite , analytique , si le symbole principal est inversible à gauche ou à droite .

Preuve du Théorème 1.1 Soit P un opérateur analytique elliptique sur une variété analytique réelle V ; et soit Q une parametrix analytique de P . Alors si T est une hyper-distribution à support compact , et si $\varphi \in C_0^\infty(V)$ est égale à 1 au voisinage du support de T , $P \varphi Q(T) - T$ est analytique au voisinage du support de T .

Si maintenant P(T) est analytique au voisinage d'un point x (qu'on supposera dans le support de T) , l'hyper-distribution $\varphi P(T)$ est bien définie , et est analytique au voisinage de x . Donc $Q \varphi P(T)$ est analytique au voisinage de x , et finalement T elle même est analytique au voisinage de x .

CHAPITRE II Opérateurs Pseudo Différentiels sur une Variété à Bord

§ 1- Opérateurs pseudo-différentiels

Soit Ω une variété à bord C^{∞}, $\partial\Omega$ son bord. On suppose plongée dans une variété ambiante V. Au voisinage d'un point du bord, on choisira toujours les coordonées locales $(x_1, \ldots x_n) = (x', x_n)$ de façon que Ω soit représenté par le demi-espace $x_n \geqslant 0$.

On note (ξ', ξ_n) la variable duale de (x', x_n)

On désigne par $\mathcal{C}_o^{\infty}(\overline{\Omega})$ (resp. $\mathcal{C}^{\infty}(\overline{\Omega})$) l'espace des fonctions C^{∞} sur Ω, à support compact dans $\overline{\Omega}$ (resp. sans condition de support), dont toutes les dérivées ont une limite sur le bord $\partial\Omega$.

Et $\mathcal{C}_o^{\infty}(\Omega)$ (resp. $\mathcal{C}^{\infty}(\Omega)$) désignent comme toujours l'espace des fonctions $\mathcal{C}^{\infty}$ à support compact (resp. sans condition de support) sur l' intérieur de Ω.

Soit P un opérateur pseudo-différentiel défini sur la variété ambiante V. Dans un système de coordonées locales, le symbole de P par la série formelle

$$\sigma(P) = \sum p_k(x, \xi)$$

où p_k est homogène de degré d-k en (si P est de degré d)

Définition 2.1 a- On dit que P est de type O s'il satisfait à la condition suivante : pour tout k, la fonction

$$(2.1) \qquad p_k(x, \xi) - e^{-(d-k)i\pi} p_k(x, -\xi)$$

s'annulle ainsi que toutes ses dérivées le long du fibré cotangent normal intérieur $(x \in \partial\Omega, \ \xi' = 0, \ \xi_n > 0)$

b- On dit que P est fortement de type O si pour tout k, la fonction définie dans (2.1) est identiquement nulle.

P est fortement de type O si et seulement si son degré d est entier , et si pour tout k , la fonction p_k a même symétrie en ξ qu'une fraction rationelle de même degré .

Un opérateur pseudo-différentiel analytique de type O est fortement de type O (du moins si la variété à bord Ω est connexe) (parcequ'une fonction analytique sur un domaine connexe est nulle si toutes ses dérivées s'annullent simultanément quelque part)

Exemple : un opérateur différentiel , la parametrix d'un opérateur différentiel elliptique sont fortement de type O .

On a alors le résultat suivant : si $f \in C_0^\infty(\bar{\Omega})$, on note $\tilde{f}$ le prolongement de f par O dans $V - \Omega$.

On définit l'opérateur P_Ω : $C_0^\infty(\bar{\Omega}) \to C^\infty(\Omega)$ par

(2.2) $P_\Omega(f) = P(\tilde{f}) / \Omega$

Théorème 2.2 Si P est de type O , P_Ω est continu : $C_0^\infty(\bar{\Omega}) \to C^\infty(\bar{\Omega})$. (ie. si $f \in C_0^\infty(\bar{\Omega})$, toutes les dérivées de $P_\Omega f$ ont une limite le long du bord $\partial\Omega$)

Si de plus P est un opérateur pseudo-différentiel analytique et si f est analytique au voisinage d'un point du bord , il en est de même de $P_\Omega f$

Nous donnons une indication de la preuve du théorème dans le cas où Ω est le demi-espace $x_n \geqslant 0$, plongé dans $V = R^n$ et où P est un opérateur de convolution par une distribution pseudo-homogène . (C'est le cas important ; et le cas général s'en déduit sans grande difficulté) .

Pour commencer , rappelons comment se comporte la transformée de Fourier - Laplace d'une fonction $f \in C_0^\infty(R^+)$:

Lemme Si $f \in C_c^\infty(\mathbb{R}^+)$, et si $\tilde{f}$ désigne le prolongement de f par O pour $x < O$, la transformée de Fourier de $\tilde{f}$:

$$\hat{f}(\xi) = \int e^{-ix.\xi} f(x)\, dx$$

(i) est holomorphe pour $\mathrm{im}\, \xi < O$

(ii) et admet pour $\xi \to \infty$, $\mathrm{im}\,\xi \leq O$, le développement asymptotique :

$$f(\xi) \sim \sum f^{(n)}(O)\quad (i\xi)^{-n-1}$$

(i) est évident - en fait il suffit que f ne croisse pas trop vite à l'infini pour que l'intégrale converge pour $\mathrm{im}\,\xi \leq O$.

Pour prouver (ii), écrivons

$$f = \sum_{n \leq N-1} f^{(n)}(0)\quad x_+^n \,/\, n! \qquad + \quad g$$

On en déduit

$$\hat{f} = \sum_{n \leq N-1} f^{(n)}(0) \quad (i\xi)^{-n-1} \qquad + \quad \hat{g}$$

En outre la dérivée N-ième de g est une fonction bornée à support compact. Par suite,

$$(i\xi)^N \hat{g} = \int e^{-ix.} \; g^{(N)}(x)\, dx$$

tend vers 0 quand $\xi \to \infty$, $\mathrm{im}\,\xi \leq 0$

Revenons maintenant à la preuve du théorème 2.2 P est défini par

$$(2.3) \qquad \widehat{Pf} = p(\xi) \quad \hat{f}(\xi)$$

En écrivant un développement de Taylor de $p(\xi)$ pour $\xi' \to 0$ et en utilisant l'homogénéité de P, on voit que pour $\xi_n \to \pm\infty$ p admet le développement asymptotique

$$(2.4) \quad p(\xi', \xi_n) \sim \sum_q a_q^{\pm}(\xi') \quad \xi_n^{d-q}$$

où $a_q^{\pm}$ est un polynôme de degré q de ξ'.

La condition P est de type 0 signifie exactement qu'on a

$$a_q^-(\xi') = e^{(d-q)i\pi} \; a_q^+(\xi')$$

Si bien qu'on peut reécrire (1.4) sous la forme

$$(1.5) \quad p(\xi', \xi_n) \sim \sum a_q(\xi') \; \xi_n^{d-q}$$

où la détermination de ξ_n^{d-q} choisie est celle qui se prolonge analytiquement pour $\operatorname{im} \xi_n \geqslant 0$

On suppose maintenant $f \in C_0^\infty(\overline{R}^n_+)$. Du lemme résulte que f admet le développement asymptotique :

$$f \sim \sum f_q(\xi') \; \xi_n^{-q-1}$$

(où f_q est, à un facteur près, la transformée de Fourier en x' de $(\partial/\partial x_n)^q f(x', 0)$)

Il en résulte que $p(\xi) \hat{f}(\xi)$ admet un développement asymptotique de la forme :

$$p(\xi) \; \hat{f}(\xi) \sim \sum b_q(\xi') \; \xi_n^{d-q-1} \qquad (\xi_n \to \infty)$$

(où comme ci-dessus, la détermination de ξ_n^s choisie est celle qui se prolonge analytiquement pour $\operatorname{im}\xi_n \geqslant 0$)

Finalement, pour tout N, on a une décomposition :

$$p(\xi) \; \hat{f}(\xi) = \sum_{q \leq N-1} b_q(\xi') \; pf(\xi_n^{d-q-1}) + r_N(\xi)$$

D'où

$$Pf = \sum_{q \leq N-1} \overline{\mathcal{F}}(b_q(\xi') \; pf(\xi_n^{d-q-1})) + \overline{\mathcal{F}} r_N(\xi)$$

(ici $\overline{\mathcal{F}}$ désigne l'opérateur transformation de Fourier inverse, et et $pf(\xi_n^s)$ désigne la distribution partie finie associée à ξ_n^s)

Or $\overline{\mathcal{F}}(b_q(\xi') \; pf(\xi_n^{d-q-1}))$ est nulle pour $x_n > 0$, parceque ξ_n^{d-q-1} se prolonge analytiquement pour $\operatorname{im} \xi_n \geqslant 0$ (ou, éventuellement, si d-q-1 est entier négatif, et si on a mal choisi la partie finie, c'est on polynôme de x_n pour $x_n \geqslant 0$)

Donc pour $x_n > 0$, $P_\Omega f$ égale la transformée de Fourier de r_N (éventuellement, modulo un polynôme). Enfin on voit sans peine que r_N est de l'ordre de $|\xi|^{d-N-1}$ pour $\xi \to \infty$. Ainsi plus N est grand, plus la transformée de Fourier de r_N admet de dérivées continues . Il s'en suit bien que P f est indéfiniment dérivable pour $x_n \geqslant 0$ (jusqu'au bord)

Pour démontrer la deuxième partie du théorème 2.2 , on procède de la même façon ; il faut en outre donner une majoration précise de r_N , qui permet de majorer les dérivées successives de $P_{\Omega} f$ et de conclure à con analyticité . On trouvera un e démonstration précise dans (4)

§ 2- Noyaux de Poisson

Ces opérateurs ont été décrits dans (3) . Nous donnons ici une description un peu différente , en particulier pour le symbole . En outre nous nous limitons aux opérateurs qui conservent localement l'analyticité : cela simplifie la définition (on a affaire à des développements en série convergentes , au lieu de développements asymptotiques) ;et le calcul symbolique est aussi un peu plus simple .

En gros , les noyaux de Poisson servent à construire des "potentiels" - par exemple à exprimer la solution d'une équation elliptique $P(f) = 0$ en fonction de sa trace sur le bord et des traces des dérivées successives .

Exemple : l'opérateur qui résout le problème de Dirichlet dans le demi-espace :

$$\begin{cases} \Delta F = 0 & \text{dans le demi-espace } R^n_+ \\ F = f & \text{sur le bord } R^{n-1} \end{cases}$$

est donné par la formule :

$$(2.6) \qquad F = (2\pi)^{-n+1} \int e^{ix'.\xi'} \, e^{-x_n.|\xi'|} \, \hat{f}(\xi') \, d\xi'$$

ou encore

$$(2.7) \qquad F = (2\pi)^{-n} \int e^{ix.\xi} \, (|\xi'| + i\xi_n) \, \hat{f}(\xi') \, d\xi$$

(la deuxième intégrale est une valeur principale - plus exactement c'est

$$\lim_{\varepsilon \to +0} \int e^{ix.\xi} \; (|\xi'| + \varepsilon + i\xi_n)^{-1} \; \hat{f}(\xi') \; d\xi \qquad)$$

La fonction $(|\xi'| + i\xi_n)^{-1}$ qui intervient sous le signe somme a les propriétés suivantes :

elle est homogène de degré -1 ;

elle est holomorphe nulle à l'infini en ξ_n dans le domaine complexe

$$|\xi_n - i|\xi'|| > (1 - \varepsilon)\,|\xi'|$$

Sa transformée de Fourier partielle inverse par rapport à ξ_n :

$$e^{-x_n.|\xi'|} = \bar{\mathcal{F}}_{\xi_n} \; (|\xi'| + i\,\xi_n)^{-1}$$

est analytique pour $x_n \geq 0$, à décroissance exponentielle pour $x_n \to \infty$ ou $\xi' \to \infty$

Nous donnons maintenant la définition d'un noyau de Poisson analytique sur un ouvert Ω du demi-espace R^n_+ , de bord $\partial\Omega = \bar{\Omega} \cap R^{n-1}$

<u>Définition 2.3</u> <u>Un opérateur linéaire continu</u> $K : C^\infty_0(\partial\Omega) \to C^\infty(\Omega)$ <u>est un noyau de Poisson analytique si son noyau-distribution</u> $K(x, y')$ <u>vérifie les conditions suivantes</u> :

(i) <u>il est analytique en dehors de la diagonale de</u> $\partial\Omega$.

(ii) <u>au voisinage de la diagonale , il admet la décomposition</u> :

(2.8) $$K(x, y') = R(x, y') + \sum K_p(x', x'-y', x_n)$$

<u>où</u> a- <u>la fonction</u> $R(x, y')$ <u>est analytique</u>

b- $K_p(x', z)$ <u>est pseudo-homogène en</u> z (1); <u>et si</u> $k_p(x', \xi)$ <u>désigne sa transformée de Fourier par rapport à</u> z , <u>pour tout compact</u> $X \subset \partial\Omega$ <u>il existe des constantes</u> ε , $R > 0$ <u>(indépendantes de</u> p <u>) telles que</u> $k_p(x', \xi)$ <u>soit holomorphe , nulle pour</u> $\xi_n = \infty$, <u>dans le domaine complexe</u> :

(1) de degré $-n-d+1+p$

$$d(x',X) \leq \varepsilon \quad , \quad |\mathrm{im}\, \xi'| < \varepsilon |\mathrm{re}\, \xi'| \quad , \quad |\xi_n - iR|\xi'|| > (R-\varepsilon)|\xi'|$$

c- Dans la formule (2.8), la série $\sum K_p(x', x'-y', x_n)$ converge dans un domaine complexe de la forme

$$d(x', X) < \varepsilon \quad ; \quad |x'-y'| + |x_n| < \varepsilon$$

$$|x'-y'| > -\varepsilon \, \mathrm{re}\, x_n \quad ;$$

$$|\mathrm{im}\; x'-y'| + |\mathrm{im}\, x_n| < \varepsilon \, (|x'-y'| + |x_n|)$$

La condition b- implique aussi que la transformée de Fourier inverse partielle $k_p(x', \xi', x_n)$ de $k_p(x', \xi)$ par rapport à ξ_n est analytique pour $x_n \geq 0$, et à décroissance exponentielle pour $\xi' \to \infty$ ou $x_n \to \infty$. (Nous renvoyons à [4] pour une démonstration plus détaillée).

Comme au chapitre I , on constate que la définition 2.3 est invariante par changement de coordonnées analytique . Ceci permet de définir un noyau de Poisson analytique sur une variété à bord analytique réelle .

Le symbole complet de K est la classe de K modulo les opérateurs à noyau-distribution analytique ; il est complètement déterminé par la série formelle :

$$(2.9) \qquad \sum k_p(x', \xi)$$

Le symbole principal de K est la forme différentielle

$$(2.10) \qquad \sigma_o(K) = k_o(x', \xi) \; d\xi_n$$

(cette définition est justifiée par le fait que dans un

changement de coordonnées analytique , le premier terme $k_o(x', \xi)$ se transforme comme la densité d'une forme différentielle en ξ_n)

Le degré de K est le degré de son symbole principal (c'est à dire $1 + \deg_\xi k_o(x', \xi)$)

Un tel opérateur jouit de la propriété suivante , qui se démontre comme au chapitre I (nous renvoyons à [4] pour une démonstrat. détaillée)

Proposition 2.4 Si T est une hyper fonction à support compact sur $\partial\Omega$, K(T) est bien définie ; et c'est une fonction analytique à l'intérieur de Ω .

Si de plus T est analytique au voisinage d'un point x du bord , il en est de même de K(T) (autrement dit K(T) admet un développement en série de Taylor convergent au voisinage de x)

En outre c un noyau de Poisson analytique est en particulier un noyau de Poisson au sens de [3] . En particulier on a

Proposition 2.5 Si K est un noyau de Poisson , il se prolonge en un opérateur continu $H^s_{comp}(\partial\Omega) \longrightarrow H^{s-d+1/2}_{loc}(\Omega)$

(H^s_{comp} désigne l'espace des fonctions à support compact qui sont localement dans H^s ; H^s_{loc} désigne l'espace de toutes les fonctions qui sont localement dans H^s)

Preuve Nous nous limiterons au cas où Ω est le demi-espace R^n_+ , de bord $\partial\Omega = R^{n-1}$, et où K est un opérateur "de convolution" - plus précisément , nous supposerons que K est défini par la formule :

$$\widehat{Kf} = k(\xi)\, \varphi(\xi')\, \hat{f}(\xi') \tag{2.11}$$

où $\varphi(\xi')$ est continue , nulle pour $|\xi'| < 1$, égale à 1 pour $|\xi'| > 2$;

$k(\xi)$ est continue pour $\xi' \neq 0$, holomorphe nulle à l'infini en ξ_n dans le domaine complexe $(\xi_n - iR|\xi'|) > (R - \varepsilon)|\xi'|$

et vérifie dans ce domaine l'inégalité :

(2.12) $\quad |k(\xi)| \leqslant c\ (1+|\xi'|)^{d}\ (1+|\xi|)^{-1} \qquad$ pour $|\xi'|>1$

premier cas si $d = 1/2$, l'opérateur K est continu : de $L^2(\partial\Omega)$ dans $L^2(\Omega)$:

cela résulte immédiatement du fait que, à cause de l'inégalité (2.12), l'intégrale $\int |k(\xi)|^2\, d\xi_n$ est bornée indépendemment de ξ' pour $|\xi'|>1$.

cas général remarquons d'abord que l'opérateur Λ^{-s}, défini sur $\partial\Omega$ par

$$\widehat{\Lambda^{-s} f}(\xi') = (1+|\xi'|^2)^{-s/2}\ \hat{f}(\xi')$$

est un isomorphisme de $L^2(\partial\Omega)$ sur $H^s(\partial\Omega)$

D'autre part, l'opérateur Λ_+^t défini sur R^n par

$$\widehat{\Lambda_+^t f}(\xi) = (1+|\xi'| - i\xi_n)^t\ \hat{f}(\xi)$$

est un isomorphisme de $H^t(R^n)$ sur $L^2(R^n)$

En outre, parceque la fonction $(1+|\xi'| \pm i\xi_n)^t$ est holomorphe pour $\operatorname{im} \xi_n \geqslant 0$, Λ_+^t laisse invariant l'espace des fonctions à support dans le demi-espace négatif $x_n \leqslant 0$.

Par suite, Λ_+^t induit un isomorphisme de $H^t(R_+^n)$ sur $L^2(R_+^n)$

Finalement il suffira de prouver que l'opérateur

$$K'' = \Lambda_+^{s-d+1/2}\ K\ \Lambda^{-s}$$

est continu de $L^2(R^{n-1})$ dans $L^2(R_+^n)$

Or on a $\widehat{K''f} = k''(\xi)\ \varphi(\xi')\ \hat{f}(\xi')$

où k'' est défini par les conditions :

(i) k'' est holomorphe, nulle à l'infini en ξ_n pour $\operatorname{im} \xi_n < 0$

(ii) $(1+|\xi'| + i\xi_n)^{s-d+1/2}\ k(\xi)\ (1+|\xi'|^2)^{-s/2} - k''$

est holomorphe pour $\operatorname{im} \xi_n \geqslant 0$.

c'est à dire

$$k''(\xi', \xi_n) = -1/2i\pi \int_{\Gamma} a(\xi', t) \, dt \,/\, t-\xi_n$$

où $a(\xi)$ désigne la fonction $(1+|\xi'| + i\xi_n)^{s-d+1/2} k(\xi) \, (1+|\xi|^2)^{-s/2}$
et Γ désigne le cercle $\xi_n - iR|\xi'| = (R-\varepsilon)|\xi'|$, orienté dans le sens positif .

Il est clair avec cette formule que k'' vérifie les mêmes hypothèses que k , le degré d étant remplacé par $1/2$. D'où le résultat .

(Remarque Λ_+^t n'est pas un opérateur pseudo-différentiel , du moins au sens de Kohn et Nirenberg , parceque pour $\xi' = 0$,son symbole n'est pas assez dérivable . La démonstration ci-dessus montre cependant que cet opérateur se compose bien avec un noyau de Poisson)

§ 3- Autres opérateurs

Pour pouvoir construire une parametrix à un problème aux limites elliptique pseudo-différentiel , nous aurons encore besoin de deux espèces d'opérateurs .

Les premiers sont les opérateurs trace analytiques , que nous construisons en deux étapes :

D'abord l'opérateur T : $C_0^{\infty}(\Omega) \to C^{\infty}(\partial\Omega)$ défini par

$$T(f) = \sum_{j \leq p-1} Q_j \left((\partial/\partial x_n)^j f(x', 0) \right) \tag{2.13}$$

où Q_j est un opérateur pseudo-différentiel analytique sur le bord , $(\partial/\partial x_n)$ désigne la n-ième dérivation partielle dans le cas où Ω est contenu dans le demi-espace R^n_+ , ou n'importe quel champ de vecteurs analytique transverse au bord dans le cas général ;
et $f(x', 0)$ désigne la restriction de f au bord

A cela , on ajoute les opérateurs qui sont adjoint d'un noyau de Poisson analytique

(2.14) $(Tf \mid g) = (f \mid Kg)$

On appelle opérateur trace analytique de classe 0 l'opérateur défini par la formule (2.14) ; et opérateur trace analytique de classe p la somme des opérateurs (2.13) et (2.14)

Le symbole principal de T est défini comme suit :

le symbole principal de l'opérateur $f \to Q((\partial/\partial x_n)^j f(x', 0))$ est

(2.15) $t(x', \xi) = q_o(x', \xi') \quad (i\xi_n)^j$

le symbole principal de l'adjoint de K est

(2.16) $t_o(x', \xi) = k_o(x', \xi', -\xi_n)$

si le symbole de K est $k_o(x', \xi', \xi_n)\, d\xi_n$

Ainsi si T est un opérateur trace de classe p , son symbole principal $\sigma_o(T)$ est une fonction sur la restriction au bord du fibré cotangent à Ω , holomorphe en ξ_n pour $\operatorname{im} \xi_n \geqslant 0$, méromorphe à l'infini , avec un pôle d'ordre au plus p-1 .

Le degré de T est le degré d'homogénéité de son symbole principal .

Le symbole complet de T est la classe de T modulo les opérateurs "à noyau analytique" (cf. ci-dessous formule (2.17)) Comme pour les opérateurs pseudo-différentiels et les noyaux de Poisson , on peut le représenter (une fois choisi un système de coordonnées) par une série formelle de fonctions sur la restriction à $\partial\Omega$ du fibré cotangent à Ω :

$$\sum t_k(x', \xi)$$

(si T est de classe p , toutes ces fonctions sont holomorphes en ξ_n pour $\operatorname{im} \xi_n \geqslant 0$, méromorphes à l'infini , avec un pole d'ordre au plus p-1)

Deux opérateurs trace de classe p qui ont même symbole diffèrent par un opérateur de la forme :

$$(2.17) \quad f \to \int_{\Omega} r(x',y)\, f(y)\, dy \quad + \sum_{j \leq p-1} \int_{\partial\Omega} r_j(x',y')\, f_j(y')\, dy'$$

où f_j désigne la j-ième dérivée normale de f ; et $r(x',y)$ et les $r_j(x',y')$ sont analytiques sur $\partial\Omega \times \bar{\Omega}$ (resp. $\partial\Omega \times \partial\Omega$)

Ces opérateurs jouissent des propriétés suivantes , qui résultent immédiatement des propriétés des noyaux de Poisson , et de celles des opérateurs pseudo-différentiels :

Théorème 2.6 Soit T un opérateur Trace analytique de classe p , de degré d . Alors

1 T se prolonge continument $H^s_{comp}(\bar{\Omega}) \to H^{s-d-1/2}_{loc}(\partial\Omega)$ pour $s-p-1/2 > 0$

2 Si f est analytique au voisinage d'un point x du bord ; il en est de même de T (f) .

Si de plus T est de classe O ,

3 Il est continu $H^s_{comp}(\bar{\Omega}) \to H^{s-d-1/2}_{loc}(\partial\Omega)$ pour tout s .

4 Si f est une hyper-distribution à support compact sur $\bar{\Omega}$, T(f) est bien définie ; c'est une hyper-distribution sur $\partial\Omega$, analytique en dehors de l'intersection de $\partial\Omega$ et du support de T .

Nous aurons enfin besoin d'un dernier type d'opérateurs : les noyaux de Green singuliers (analytiques) , qui généralisent en gros le composé $K \circ T$ d'un noyau de Poisson et d'un opérateur trace .

Nous nous contentons ici de décrire les noyaux de Green singuliers "homogènes , à coefficients constants " , sur le demi-espace $\mathbb{R}^n_+$. Le cas général se construit à partir de celui-là , à coup de développements en series convergentes , exactement comme pour les noyaux de Poisson , ou les opérateurs pseudo-différentiels analytiques

Comme pour les opérateurs trace, nous faisons la construction en deux étapes

D'abord l'opérateur

(2.18) $G(f) = K\left((\partial/\partial x_n)^j f(x', 0)\right)$

est appelé noyau de Green singulier de classe p si $j \leq p-1$, si K est un noyau de Poisson analytique.

Son symbole principal est

(2.19) $\sigma_o(G) = g_o(x', \xi', \xi_n, \eta_n)\, d\xi_n =$

$= k_o(x', \xi', \xi_n) \quad (i\eta_n)^j \quad d\xi_n$

si K admet pour symbole principal $k_o(x', \xi', \xi_n) \quad d\xi_n$

Le deuxième type, appellé noyau de Green singulier de classe O, est l'opérateur $f \to G(f)$ défini par :

(2.20) $\widehat{Gf} = (2\pi)^{-1} \int_{-\infty}^{+\infty} g(\xi', \xi_n, \eta_n)\, \hat{f}(\xi', \eta_n)\, d\eta_n$

où la fonction $g(\xi', \xi_n, \eta_n)$ a les propriétés suivantes :

a. elle est homogène de degré d-1 en (ξ', ξ_n, η_n).

b. elle est holomorphe, nulle pour $\xi_n = \infty$ ou $\eta_n = \infty$ dans le domaine complexe

$|\mathrm{im}\, \xi'| < \varepsilon |\mathrm{re}\, \xi'|$

$|\xi_n - iR|\xi'|| > \varepsilon(R - \varepsilon)|\xi'|$ (en particulier pour $\mathrm{im}\, \xi_n \leq 0$)

$|\eta_n + iR|\xi'|| > \varepsilon(R - \varepsilon)|\xi'|$ (en particulier pour $\mathrm{im}\, \eta_n \geq 0$)

(Nous renvoyons à [4] pour la définition détaillée dans le cas général)

Le symbole principal de G est

(2.21) $\sigma_o(G) = g(x', \xi', \xi_n, \eta_n) \quad d\xi_n$

(ici, bien sur, il ne dépend pas de x').

Enfin on appelle noyau de Green singulier de classe p la somme des opérateurs définis par les formules (2.18) et (2.20)

Comme pour les noyaux de Poisson ,les opérateurs trace , etc.. , on définit le symbole complet d' un noyau de Green singulier G comme la classe de G modulo les opérateurs "à noyau-distribution analytique ". Et on peut représenter ce symbole par une série formelle de formes différentielles en ξ_n du type (2.21) . Deux noyaux de Green singuliers de classe p qui ont même symbole diffèrent par un opérateur de la forme :

(2.22) $$f \rightarrow \int_{\Omega} r(x,y)\, f(y)\, dy \quad + \quad \sum_{j \leqslant p-1} \int_{\partial\Omega} r_j(x,y')\, f_j(y')\, dy'$$

(où comme ci-dessus , f_j désigne la restriction au bord de la j-ième dérivée normale de f , et les fonctions r(x,y) et $r_j(x,y')$ sont analytiques sur $\bar{\Omega} \times \bar{\Omega}$ (resp. $\partial\Omega \times \bar{\Omega}$))

Ces opérateurs jouissent des propriétés suivantes , qui généralisent celles du composé d'un noyau de Poisson et d'un opérateur trace (nous renvoyons à [4] pour la preuve)

Théorème 2.7 Si G est un noyau de Green singulier analytique de classe p , de degré d

1) G se prolonge continument $H^s_{comp}(\bar{\Omega}) \rightarrow H^{s-d}_{loc}(\bar{\Omega})$

pour $s-p-1/2 > 0$

2) Pour toute f , G(f) est analytique à l'intérieur de Ω .

3) Si f est analytique au voisinage d'un point x du bord , il en est de même de G(f) .

Si de plus G est de classe O

4) G est continu $H^s_{comp}(\bar{\Omega}) \rightarrow H^{s-d}_{loc}(\bar{\Omega})$ pour tout s .

5) Et si f est une hyper-distribution à support compact dans $\bar{\Omega}$ G(f) est bien définie ; c'est une fonction analytique dans l'intérieur de Ω

§4- Composition et calcul symbolique

Comme au chapitre 1, on ne peut pas composer tels quels nos opérateurs. On a seulement un résultat local, qu'on obtient en tronquant comme au chapitre I

Le théorème suivant donne un tableau des résultats. Comme au début du chapitre, Ω désigne une variété à bord analytique réelle, de bord $\partial\Omega$. Ω' désigne une sous variété ouverte de Ω, de bord $\partial\Omega' \subset \partial\Omega$; relativement compacte dans $\bar{\Omega}$; et φ désigne une fonction C^∞ à support compact dans $\bar{\Omega}$, égale à 1 dans Ω'.

Théorème 2.8 On désigne génériquement par

- P un opérateur pseudo-différentiel analytique de type 0 sur $\bar{\Omega}$
- Q un opérateur pseudo-différentiel analytique sur $\partial\Omega$
- K un noyau de Poisson analytique sur Ω
- T un opérateur trace analytique sur Ω
- G un noyau de Green singulier analytique sur Ω

Enfin x_n désigne une fonction analytique au voisinage de $\partial\Omega$, nulle à l'ordre 1 exactement sur $\partial\Omega$ (la fonction n-ième coordonnée si Ω est un ouvert du demi-espace R^n_+) ; δ désigne une mesure à densité analytique sur $\partial\Omega$ (la mesure de Haar usuelle de R^{n-1} si Ω est un ouvert du demi-espace R^n_+)

Alors

1) $P\varphi K$, $G\varphi K$, $K\varphi Q$, et l'opérateur $f \to P(f\cdot\delta)/\Omega$ sont des noyaux de Poisson analytiques sur Ω'

2) $T\varphi P$, $T\varphi G$, $Q\varphi T$ sont des opérateurs trace analytiques sur Ω'

3) $T\varphi K$ est un opérateur pseudo-différentiel analytique sur $\partial\Omega'$.

4) $K\varphi T$, $P\varphi G$, $G\varphi P$, $G^1\varphi G^2$, $P^1_\Omega\varphi P^2_\Omega - (P^1\varphi P^2)_\Omega$ sont des noyaux de Green singuliers analytiques sur Ω'

5) $x_n^p K$ est un noyau de Poisson analytique de degré $\deg K - p$

$T x_n^q$ est un opérateur trace analytique de degré $\deg T - q$

$x_n^p \, G \, x_n^q$ est un noyau de Green singulier analytique de degré $\deg G - p - q$

Ces résultats se prouvent tous comme au chapitre I : pour etudier le composé $A \varphi B$, on examine d'abord le cas où B est l'opérateur de multiplication par une fonction analytique ; puis le cas où B est un opérateur "à coefficients constants", sur le demi-espace R_+^n, et où A et B admettent tous deux une représentation par intégrale de Fourier du même genre qu'au chapitre I . Le cas général en résulte par un argument de produits tensoriels topologiques . Nous ne donnerons pas de preuve détaillée ici , et renvoyons à [4] pour cela .

Nous donnons maintenant le symbole principal de tous ces composés (il y a aussi une formule pour le symbole complet , obtenue en gros en combinant la formule pour le symbole principal et la formule de Leibnitz , comme pour les opérateurs pseudo-différentiels - cf. [4] , formules (6.10) à (6.14))

Pour décrire le symbole des composés nous aurons besoin de opérations suivantes :

Soient f(z) et g(z) deux fonctions d'une variable réelle , analytiques et méromorphes à l'infini .

Soit Γ un cercle $|z - iR| = R - \varepsilon$, tel que f et g soient holomorphes dans un voisinage de l'ensemble $\operatorname{im} z \geqslant 0$, $|z - iR| > R - \varepsilon$. On oriente Γ positivement.

On pose

(2.23) $h^+(f) = -1/2i\pi \int_\Gamma f(t) \, dt/t-z$

(2.24) $H^-(f) = f - h^+(f)$

(2.25) $\int^+ f\, g = \int^+ f(t)\, g(t)\, dt = - \int_\Gamma f(z)\, g(z)\, dz$

Ces définitions ne dépendent pas du choix de Γ ; ainsi $h^+(f)$ est l'unique fonction holomorphe , nulle à l'infini , au voisinage de la demi sphère de Riemann $\operatorname{im} z \leqslant 0$, telle que $f - h^+(f)$ soit holomorphe dans le demi-plan $\operatorname{im} z > 0$.

Et si f et g sont nulles à l'infini , on a $\int^+ f\, g = \int_{-\infty}^{+\infty} f(t)\, g(t)\, dt$

On a alors les résultats suivants (nous omettons dans les formules qui suivent la fonction φ qui sert à tronquer les opérateurs à composer)

Théorème 2.9 Si p_o , q_o , k_o , t_o , g_o désignent respectivement les symboles proncipaux de P , Q , K , T , G , on a :

1) $\sigma_o(P\ K) = h^+_{\xi_n}(\, p_o(x',\xi',\xi_n)\, k_o(x',\xi',\xi_n)\,)\ d\xi_n$

$\sigma_o(G\ K) = (\int^+ g_o(x',\xi',\xi_n,\eta_n)\, k_o(x',\xi',\eta_n)\, d\eta_n\ )\ d\xi_n$

$\sigma_o(K\ Q) = k_o(x',\xi',\xi_n)\ q_o(x',\xi')\ d\xi_n$

$\sigma_o(\, f \to P\ (f\delta)\ ) = h^+_{\xi_n}\, p_o(x',\xi',\xi_n)\ d\xi_n$

(dans ce dernier cas , on suppose que Ω est ouvert dans le demi-espace $\mathbb{R}^n_+$, et que δ est la mesure de Haar usuelle du bord $\mathbb{R}^{n-1}$)

2) $\sigma_o(T\ P\ ) = H^-(\, t_o\ .\ p_o\ )$

$\sigma_o(T\ G) = \int^+ t_o(x',\xi',t)\ g_o(x',\xi',t,\xi_n)\ dt$

$\sigma_o(Q\ T) = q_o\ .\ t_o$

3) $\sigma_o\ (T\ K) = \int^+ t_o(x',\xi',\xi_n)\ k_o(x',\xi',\xi_n)\ d\xi_n$

4) $\sigma_o(\, K\ T\ ) = k_o(x',\xi',\xi_n)\ \ t_o(x',\xi',\eta_n)\ \ d\xi_n$

$\sigma_o\,(P\ \ G) = h^+_{\xi_n}(\, p_o(x',\xi',\xi_n)\ \ g_o(x',\xi',\xi_n,\eta_n)\ )\ d\xi_n$

$\sigma_o\,(G\ \ P\ ) = H^-_{\eta_n}(\, g_o(x',\xi',\xi_n,\eta_n)\ \ p_o(x',\xi',\eta_n)\ )\ d\xi_n$

$$\sigma_o(G^1 G^2) = \left(\int^+ g_o^1(x', \xi', \xi_n, t)\; g_o^2(x', \xi', t, \eta_n)\, dt \right) d\xi_n$$

$$\sigma_o((P^1P^2)_\Omega - P^1_\Omega P^2_\Omega) =$$

$$h^+_{\xi_n} H^-_{\eta_n}\left[(p^1_+(\xi_n) - p^1_+(\eta_n))\,(p^2_-(\xi_n) - p^2_-(\eta_n)) / i(\xi_n - \eta_n) \right] d\xi_n$$

(dans la dernière formule , on a posé

$$p^1_+(\xi_n) = h^+_{\xi_n}(p^1_o(x', \xi', \xi_n))$$

$$p^2_-(\xi_n) = H^-_{\xi_n}(p^2_o(x', \xi', \xi_n)) \qquad)$$

5) $\sigma_o(x_n^p K) = (i\,\partial/\partial\xi_n)^p\, k_o(x', \xi', \xi_n) \quad d\xi_n$

$$\sigma_o(T x_n^q) = (-i\,\partial/\partial\xi_n)^q\, t_o(x', \xi', \xi_n)$$

$$\sigma_o(x_n^p G x_n^q) = (i\,\partial/\partial\xi_n)^p (-i\,\partial/\partial\eta_n)^q\, g_o(x', \xi', \xi_n, \eta_n) \quad d\xi_n$$

(dans ces trois dernières formules , il faut que la différentielle dx_n soit le covecteur normal intérieur $\partial/\partial\xi_n$; c'est le cas si Ω est un ouvert de R^n_+ , x_n la dernière fonction coordonnée)

Pour tous ces résultats , le cas important est celui des opérateurs "homogènes à coefficients constants " . C'est lui qui fournit les formules pour les symboles principaux , et la première étape vers la démonstration du théorème 2.8 . Exemple :

Si P est l'opérateur pseudo-différentiel défini par

$$\widehat{Pf} = p(\xi)\; \hat{f}(\xi)$$

et si f est une fonction C^∞ à support compact sur R^{n-1} , la fonction qui vaut O pour $x_n < O$ et $P(f\delta)$ pour $x_n > O$, a pour transformée de Fourier

$$h^+_{\xi_n}(p(\xi)\; \hat{f}(\xi')) = h^+ p(\xi', \xi_n) \quad \hat{f}(\xi') \qquad (')$$

D'où la dernière formule du théorème 2.9 , n° 1) . Les autres formules sont du même ordre de difficulté , sauf celle qui concerne la différence $(P^1P^2)_\Omega - P^1_\Omega P^2_\Omega$,pour laquelle nous renvoyons à [4] .

Pour terminer , signalons le résultat suivant , qui servira au chapitre III pour la construction de parametrix :

Proposition 2.10 Si G est un noyau de Green singulier analytique de classe p , de degré -1 , il existe un noyau de Green singulier analytique de classe p , de degré -1 , G' , tel qu'on ait :

$$\sigma(1-G)(1+G') = \sigma(1)$$

Autrement dit , le symbole complet de noyau de Green :

$$\sum_{k \geqslant 1} \sigma(G)^k$$

est celui d'un noyau de Green singulier analytique . Cette proposition se prouve comme le corollaire 1.9 (chap. I) ; il est commode d'introduire une norme formelle adaptée aux formules de composition des noyaux de Green singuliers analytiques .(cf. [4] pour une preuve détaillée)

(*) Du fait que $p(\xi', \xi_n)$ est analytique , de type O , il résulte immédiatement que

$$h^+_{\xi_n}(p(\xi', \xi_n))$$

est symbole principal d'un noyau de Poisson analytique .

CHAPITRE III PROBLEMES AUX LIMITES PSEUDO DIFFERENTIELS

Remarquons tout d'abord que les définitions des chapitres I et II se généralisent sans aucune difficulté aux opérateurs (ou systèmes d'opérateurs) qui opèrent sur les sections d'un fibré analytique . c'est de tels systèmes qu'il s'agira ici . En fait la généralisation aux fibrés est indispensable pour l'énoncé du théorème 3.1

Le but du chapitre est de construire (quand cela est possible) une parametrix à gauche , ou à droite , ou bilatère , au problème suivant :

comme au chapitre II , on se donne une variété à bord analytique réelle Ω , de bord $\partial\Omega$

E et E' désignent deux fibrés vectoriels complexes (analytiques) sur $\bar{\Omega}$; F et F' désignent deux fibrés vectoriels (analytiques) complexes sur $\partial\Omega$. f (resp. g) désigne une section de E (resp. E') ; u (resp. v) désigne une section de F (resp. F')

P est un "système" d'opérateurs pseudo-différentiels analytique elliptique de type O , défini au voisinage de $\bar{\Omega}$, opérant de $C_0^\infty(E)$ dans $C^\infty(E')$

K est un système de noyaux de Poisson analytiques : $C_0^\infty(F, \partial\Omega) \to C^\infty(E')$

T est un système d'opérateurs trace analytique : $C_0^\infty(E) \to C^\infty(F')$

Q est un système d'opérateurs pseudo-différentiels analytiques : $C_0^\infty(F, \partial\Omega) \to C^\infty(F', \partial\Omega)$

on se propose d'étudier le problème :

$$(3.1)\qquad \begin{cases} P_\Omega f + K u = g \\ T f + Q u = v \end{cases}$$

1 Résultats

Il sera commode de noter le couple (f, u) comme un vecteur colonne $\binom{f}{u}$. Au problème (3.1) est alors associée la matrice :

(3.2) $$A = \begin{pmatrix} P_\Omega & K \\ T & Q \end{pmatrix}$$

On introduit aussi la notation suivante : si f est un section C^∞ de E, et u une section C^∞ de F, on notera

(3.3) $$\gamma_m \binom{f}{u}$$

la suite $(f_o, f_1, \dots f_{m-1}, u) \in C^\infty(E, \partial\Omega)^m \times C(F, \partial\Omega)$ où f_j désigne la trace sur $\partial\Omega$ de la j-ième dérivée normale de f (ceci suppose choisi une fois pour toutes un champ de vecteurs analytique transverse à $\partial\Omega$, au voisinage de $\partial\Omega$).

On commence par s'occuper de l'opérateur P tout seul. Pour simplifier l'énoncé, nous supposerons que Ω est compacte, connexe. (En fait le résultat ci-dessous provient d'un résultat de calcul symbolique, décrit au paragraphe 2)

Théorème 3.1 (On suppose Ω compacte connexe)

1) Il existe un fibré vectoriel analyrique F^1 sur $\partial\Omega$, et un noyau de Poisson analytique $K^1 : C_0^\infty(F', \partial\Omega) \to C^\infty(E', \bar{\Omega})$ tels que l'opérateur $\binom{f}{u} \to P_\Omega f + K^1 u = g$ possède un inverse à droite de la forme

(3.4) $$g \to \begin{pmatrix} P_\Omega^{-1} g & + G^o g \\ T\, g & \end{pmatrix}$$

(où P^{-1} déşigne une parametrix de P, au voisinage de $\bar{\Omega}$, G est un noyau de Green singulier analytique, T un opérateur trace analytique)

tels que en outre, le composé dans l'autre sens : $\begin{pmatrix} P_\Omega^{-1} + G^o \\ T \end{pmatrix} (P_\Omega \;\; K^1)$ soit de la forme

(3.5) $$1_{E \oplus F} - K^o \gamma_m$$

(où K^o désigne par abréviation la somme directe d'un noyau de

Poisson analytique et d'un opérateur pseudo-différentiel analytique sur le bord)

2) Alors l'opérateur

(3.6) $K^o \, \gamma_m$

est un projecteur sur $\operatorname{Ker} (P_\Omega \oplus K^1)$

et l'opérateur pseudo-différentiel sur le bord :

(3.7) $H^o = \gamma_m \, K^o$

est un projecteur sur l'espace des traces à l'ordre m des solutions de $P_\Omega \oplus K^1$

Montrons maintenant comment ce thèorème permet de vérifier très simplement si le problème (3.1) a une parametrix .

D'abord il est clair que (3.1) est équivalent au problème

(3.1) bis
$$\begin{cases} P_\Omega \, f + K u + K^1 u_1 = g \\ T f + Q u = v \\ u_1 = w \end{cases}$$

(ie. (3.1) et (3.1) bis ont simultanément une parametrix à gauche , à droite , ou bilatère)

Aussi nous supposerons désormais que dans la première équation de (3.1) , le fibré F et le noyau de Poisson K sont précisément ceux du théorème 3.1 . Le théorème 3.1 nous fournit alors un inverse à droite (3.4) pour la première équation , un noyau de Poisson K^o ((3.5) et (3.6)) permettant de décrire les solutions de la première équation , ainsi que le projecteur pseudo-différentiel H^o de (3.7)

Introduisons maintenant l'opérateur pseudo-différentiel sur le bord :

(3.8) $Q^1 = (T \oplus Q) \; K^o$

Enfin

(3.9) F_o désigne le fibré vectoriel sur $T^*\partial\Omega$, image du projecteur $\sigma_o (H^o)$ (ici T^*... représente le fibré cotangent privé de la section nulle)

On a alors le résultat suivant :

Théorème 3.2 Le problème (3.1) admet une parametrix à gauche (resp. à droite, resp. bilatère) si et seulement si le symbole principal $\sigma_o(Q^1)$ est une injection (resp. surjection, bijection) de F_o dans F' .

Voici la preuve dans le premier cas (parametrix à gauche) (les autres cas se démontrent exactement de la même façon)

a- La condition est nécessaire :

Supposons que le système (3.1) possède une parametrix à gauche, de matrice

$$(3.10)\quad B = \begin{pmatrix} P_\Omega^{-1} + G^2 & K^2 \\ T^2 & Q^2 \end{pmatrix}$$

Alors si A désigne la matrice (3.3) associée au problème (3.1), on a

$$(3.11)\quad H^o = \gamma_m K^o \sim \gamma_m B A K^o$$

(où le signe $\sim$ signifie que les opérateurs en question ont même symbole complet)

Si on pose alors

$$(3.12)\quad Q^3 = \gamma_m (K^2 \oplus Q^2)$$

on a, parceque $(P_\Omega \oplus K) K^o = 0$

$$(3.13)\quad Q^3 \; Q^1 \sim H^o$$

et en particulier

$$(3.14)\quad \sigma_o(Q^3) \; \sigma_o(Q^1) = \sigma_o(H^o)$$

ce qui prouve bien que $\sigma_o(Q^1)$ est une injection de F_o dans F' .

b- Avant de prouver la réciproque, apportons les précisions suivantes sur les opérateurs K^o, H^o, Q^1 etc.. qui interviennent dans le théorème 3.2 .

Soient $E = \oplus_1^p E_i$ et $F = \oplus_1^q F_j$ deux fibrés décomposés en somme directe.

Et soient $s = (s_o, s_1, \dots s_p)$, $t = (t_o, t_1, \dots t_q)$ deux suites de nombres .

(3.15) <u>Par définition, un opérateur pseudo-différentiel P (ou un noyau de Poisson, un opérateur trace, etc..) opérant des sections de E dans celles de F est de multi-degré (s, t) si dans sa matrice $(P_{ij})_{\substack{i=0,..p\\ j=0,..q}}$ l'opérateur P_{ij} est de degré $t_j - s_i$.</u>

Le symbole principal $\sigma_o(P)$ est alors la matrice $\sigma_o(P_{ij})$, où P_{ij} est considéré comme opérateur de degré $t_j - s_i$.

Ici l'opérateur trace γ_m prend ses valeurs dans l'espace $C^\infty(E^m \oplus F)$, où le fibré $E^m \oplus F$ est naturellement décomposé en somme directe (toujours en supposant choisi une fois pour toutes un vecteur normal) . Dans le théorème 3.1, on peut choisir K, K^o, H^o de façon que K soit de degré 0, K^o de multi-degré (s, O), et H^o de multi-degré (s, s) où $s = (0, 1, \dots m-1, 0)$.

Alors si dans (3.1) les opérateurs T, Q sont de degré 0, l'opérateur Q^1 est de multi-degré $(s, 0)$.

Et le symbole principal qui intervient dans le théorème 3.2 est celui de Q^1 en tant qu'opérateur de multi-degré $(s, 0)$

Supposons maintenant que $\sigma_o(Q^1)$ est injectif sur F_o . Alors il existe un symbole principal d'opérateur de multi-degré $(0, s)$ $\sigma_o(Q^3)$ tel que

$$\sigma_o(Q^3)\ \sigma_o(Q^1) = \sigma_o(H^o)$$

Dans ces conditions, $H^o(1 - Q^3 Q^1)$ est de symbole

principal nul en tant qu'opérateur de multi-degré (s, s) . Par suite il peut s'écrire sous la forme

(3.16) $H_o(1 - Q^3 Q^1) = H^o(1 - R)$

où R est de degré -1 .

Donc $(1 - R)$ possède une parametrix $(1 + R')$, où R' est lui aussi de degré -1 (Cor. 1.9 , chapitre I)

Et on a :

(3.17) $H^o(1 + R') H^o Q^3 Q^1 \sim H^o(1 + R')(1 - R) \sim H_o$

Alors , si on note B^1 ($g \to B^1(g)$) l'inverse à droite (3.4) du théorème 3.1 , l'opérateur

(3.18) $\binom{g}{v} \longrightarrow B^1 g + K^o(1 + R') Q^3 (v - (T \oplus Q) B^1 g)$

est une parametrix à droite du problème (3.1) .

(Remarque : il n'est pas du tout nécessaire d'utiliser des opérateurs à multi-degré dans le problème (3.1) ; on pourrait l'éviter en utilisant à la place de γ_m un opérateur trace de degré O . Cependant , dans un problème aux limites , en particulier dans ceux qui sont associés à un vrai opérateur différentiel , les opérateurs "à multi-degré" s'introduisent très souvent , de façon tout à fait naturelle , dans les conditions limites - ie. la deuxiéme équation de (3.1) -)

Pour terminer ce paragraphe, nous précisons le lien entre l'opérateur pseudo-différentiel P et le fibré F_o du théorème 3.1 :

On suppose que le fibré F et l'opérateur K de (3.1) sont ceux du thèorème 3.1 . Il leur correspond un opérateur K_o ((3.5)) et un projecteur pseudo-différentiel H_o ((3.7)) ; F_o est le fibré image de $\sigma_o(H_o)$

Soient maintenant F' un autre fibré vectoriel sur et K' un autre noyau de Poisson, auxquels le théorème 3.1 s'applique : $(P_\Omega \oplus K')$ possède une parametrix à droite B' telle que

$B'\ (P_\Omega \oplus K') \sim 1 - K'^o\ \gamma_{m'}$

A ces opérateurs est associé un nouveau "projecteur" H'^o (du moins le symbole complet de H'^o est un projecteur) ; et un nouveau fibré sur $T^*\partial\Omega$, F'_o = image de $\sigma_o(H'^o)$

Il s'agit de comparer F_o et F'_o .

1- changement de m :

on suppose F = F' , K = K' ; simplement m' > m . A part cela, on garde la même parametrix . On constate alors immédiatement que le nouveau fibré F'_o est la somme directe de F_o et de l'image réciproque (pull-back) de $E^{m'-m}$ sur $T^*\partial\Omega$.

2- nous supposons maintenant F et K différents de F' et K' mais (grâce à la conclusion du premier cas) m = m' .

Nous notons B l'inverse de $(P_\Omega \oplus K)$ (ie. $(P_\Omega \oplus K)\ B = 1_{E'}$) et B' l'inverse (ou la parametrix) de $(P_\Omega \oplus K')$

Introduisons maintenant l'opérateur

$(P_\Omega \oplus K \oplus K') = A'$

Il possède deux parametrix à droite :

$g \longrightarrow B^1 g = B g \oplus 0$

et $g \longrightarrow B^2 g$ où $B^2 g$ est donné par $\binom{f}{u'} = B' g$, u = 0

A ces deux inverses correspondent deux opérateurs :

$K^1\ \gamma_m = 1 - B^1\ A'$; $K^2\ \gamma_m = 1 - B^2\ A'$

et deux projecteurs : $H^1 = \gamma_m K^1$; $H^2 = \gamma_m K^2$

On vérifie immédiatement que dans la décomposition en somme $E^m \oplus F \oplus F'$, H^1 a pour matrice

$$\begin{pmatrix} H^o & & -\gamma_m B K' \\ C & O & 1 \end{pmatrix}$$

et si H'^o a pour matrice $\begin{pmatrix} a & b \\ c & d \end{pmatrix}$ dans la décomposition $E^m \oplus F'$ du fibré sur les sections duquel H'^o opère ; et si pour cette même décomposition , $\gamma_m B'$ a pour matrice $\begin{pmatrix} e \\ f \end{pmatrix}$, H^2 a pour matrice :

$$\begin{pmatrix} a & -eK & b \\ 0 & 1 & 0 \\ c & -fK & d \end{pmatrix}$$

De sorte que l'image de $\sigma_o(H^1)$ est isomorphe à $F_o \oplus F'$ l'image de $\sigma_o(H^2)$ est isomorphe à $F'^o \oplus F$.

Enfin il résulte immédiatement des définitions q'on a

$$K^2 \gamma_m K^1 \sim K^1 \quad ; \quad K^1 \gamma_m K^2 \sim K^2$$

d'où $H^2 H^1 \sim H^1$; $H^1 H^2 \sim H^2$

De sorte que $\sigma_o(H^2)$ réalise un isomorphisme de l'image de $\sigma_o(H^1)$ sur l'image de $\sigma_o(H^2)$.

Pour regrouper ces deux résultats , appelons groupe de Grothendieck relatif le groupe $K_{\partial\Omega}(T^*\partial\Omega)$ engendré par les fibrés sur $T^*\partial\Omega$ modulo ceux qui sont image réciproque d'un fibré sur $\partial\Omega$. Alors , <u>ce qui dépend intrinsèquement de P , c'est la classe du fibré F_o dans le groupe de Grothendieck relatif $K_{\partial\Omega}(T^*\partial\Omega)$</u>

Il résulte aussi du théorème 3.2

<u>Proposition 3.3 Pour qu'il existe un problème aux limites elliptique du type (3.1) associé à P (ie. admettant une parametrix bilatère) il est nécessaire et suffisant que la classe de F_o dans $K_{\partial\Omega}(T^*\partial\Omega)$ soit nulle</u>

(le théorème 3.2 montre que c'est nécessaire : si (3.1) est elliptique , le symbole principal $\sigma_o(Q^1)$ (où Q^1 est défini par (3.8)) réalise un isomorphisme de F_o sur l'image réciproque de F' ; et on vérifie aisément que c'est suffisant) .

2 Preuve du théorème 3.1

Nous indiquons ici la preuve du théorème 3.1 , qui est le résultat essentiel du paragraphe I . Nous allons en fait prouver un résultat de calcul symbolique (ie. le théorème 3.1 , localement , et où tous les signes = sont remplacés par le signe $\sim$ signifiant : ont même symbole complet) . Dans le cas où Ω est compacte , connexe il est aisé d'améliorer le résultat , à partir des remarques suivantes :

a- si R est un opérateur de symbole nul , de classe p , il existe un opérateur R' de symbole nul , de classe p tel que (1 + R') (1 - R) soit un projecteur sur l'image de (1 - R) (ou de $(1 - R)^N$, N grand) , ayant même symbole complet que l'identité .

b- si Ω est compacte connexe , de bord non vide , toute fonction analytique nulle à l'ordre infini sur le bord est identiquement nulle ; par suite, si X est un sous espace de dimension finie de l'espace des fonctions analytiques , il existe un projecteur sur X , de la forme $K \gamma_m$, où K est un noyau de Poisson de symbole nul , et γ_m est défini comme au paragraphe 1 .

La version symbolique et locale du théorème 3.1 ne fournira évidemment jamais de théorème global d'existence ; par contre elle donne le résultat de régularité suivant :

Proposition 3.4 Si le problème (3.1) est elliptique à gauche , et si , avec les notations de (3.1) , g et v sont analytiques au voisinage d'un point x du bord , il en est de même de f et u .

Et le théorème 3.2 (ou plutôt sa version locale) donne un moyen théorique de vérifier si un problème du type (3.1) est elliptique à gauche.

La première étape vers la preuve du théorème 3.1 est la construction des symboles principaux. Pour cela, le point technique essentiel est le lemme suivant (où en fait on résout, de façon grossière, une équation de Wiener - Hopf à une variable)

Lemme 3.5 Soit p un symbole principal d'opérateur pseudo-différentiel analytique de type 0 ; et g un symbole principal de noyau de Green singulier analytique. Alors

(i) il existe des symboles g', k', t' (respectivement d'un noyau de Green singulier, d'un noyau de Poisson, et d'un opérateur trace) tels que

$$g = \sigma_o(p\, g') + \sigma_o(k'\, t')$$

(ii) il existe des symboles g'', k'', et un nombre m tels que

$$g = \sigma_o(g''\, p) + \sigma_o(k''\, \gamma_m)$$

(les lois de composition qui interviènnent ici sont celles du théorème 2.9 ; et dans ce qui suit, nous reprenons les notations de ce théorème : (2.23), (2.24), (2.25))

Parceque p est de type O, il existe un polynôme de ξ_n : $a(\xi', \xi_n)$, de degré $d + 2m$ (où $d = \deg p$) tel que

$$p\, a^{-1} \quad |\xi|^{2m} \qquad \text{et} \qquad a^{-1}\, p \quad |\xi|^{2m}$$

soient très voisins de 1. Ceci fournit deux factorisations :

$$(3.19) \qquad p = a\, p'_-\, p'_+\, |\xi|^{2m}$$

$$(3.20) \qquad p = p''_-\, p''_+\, a\, |\xi|^{2m}$$

où $p'_{\pm}$, $p''_{\pm}$ est analytique pour $\xi' \neq 0$, homogène de degré 0 , holomorphe en ξ_n , ainsi que son inverse , pour $\pm \operatorname{im} \xi_n \leqslant 0$ (y compris pour $\xi_n = \infty$) [1]

Pour abréger , posons encore

$$(3.21) \qquad \xi_{\pm} = |\xi'| \pm i\xi_n \quad , \quad \eta_{\pm} = |\xi'| \pm i\eta_n$$

Soit maintenant g'_1 la fonction définie par

$$g'_1 = {p'_+}^{-1} \xi_+^m \; h^+_{\xi_n}({p'_-}^{-1} \xi_-^m \, a^{-1} \, g)$$

Comme $a \, p'_- \, \xi_-^{-m}$ est holomorphe pour $\operatorname{im} \xi_n > 0$, on a

$$h^+_{\xi_n}(a \, p'_- \, \xi_-^{-m} \, H^-_{\xi_n}({p'_-}^{-1} \xi_-^m \, a^{-1} \, g)) = 0$$

d'où

$$h^+_{\xi_n}(p \, g'_1) = h^+_{\xi_n}(a \, p'_- \, \xi_-^{-m} \, h^+_{\xi_n}({p'_-}^{-1} \xi_-^m \, a^{-1} \, g)) =$$

$$h^+_{\xi_n}(a \, p'_- \, \xi_-^{-m} \, \xi_-^m \, {p'_-}^{-1} \, a^{-1} \, g) = h^+(g) = g$$

Finalement , avec

$$(3.22) \qquad g' = h^+_{\xi_n}(g'_1)$$

$$\text{donc} \quad g'_1 = g' + \sum_{j \leqslant m-1} \xi_n^j \, t_j$$

où t_j est symbole principal d'un opérateur trace analytique

$$(3.23) \qquad k_j = h^+_{\xi_n} (\xi_n^j \, p)$$

on a

$$g = \sigma_0 (p \, g') + \sum_{j \leqslant m-1} \sigma_0 (k_j \, t_j)$$

d'où la première partie du lemme .

---

[1] attention que dans les formules (3.19) , (3.20) , l'ordre des facteurs a de l'importance , du moins quand il s'agit de "systèmes . Par contre , le facteur $|\xi|^{-2m}$, qui commute avec tout le reste , peut être intercalé n'importe où .

Si maintenant on pose

$$(3.24) \qquad g'' = H^-_{\eta_n}\left(g\, a^{-1}(\xi', \eta_n)\, p''^{-1}_{+}(\xi', \eta_n)\, \eta_+^m\right) \eta_-^m\, p''^{-1}_-$$

on a

$$H^-_{\eta_n}(g''\, p) = H^-_{\eta_n}\left(H^-_{\eta_n}(g\, a^{-1}\, p''^{-1}_+\, \eta_+^m)\, p''_+\, \eta_+^{-m}\, a\right) =$$

$$= g - H^-_{\eta_n}\left(h^+_{\eta_n}(g\, a^{-1}\, p''^{-1}_+\, \eta_+^m)\, p''_+\, \eta_+^{-m}\, a\right)$$

Or $p''_+\, \eta_+^{-m}\, a\,(\xi', \eta_n)$ est holomorphe pour $\operatorname{im} \eta_n \leqslant 0$, avec un pôle d'ordre $< m + d$ pour $\eta_n = \infty$

Par suite le reste

$$(3.25) \qquad H^-_{\eta_n}\left(h^+_{\eta_n}(g\, a^{-1}\, p''^{-1}_+\, \eta_+^m)\, p''_+\, \eta_+^{-m}\, a\,(\xi', \eta_n)\right)$$

est un polynôme de η_n, de degré $< m + d$, à coefficients symboles principaux de noyaux de Poisson analytiques.

D'où la deuxième partie du lemme (avec $m+d$ à la place de m)

Ce premier résultat va permettre de faire la construction suivante. Comme au début du chapitre, P désigne un "système" elliptique d'opérateurs pseudo-différentiels (analytiques) de type 0, opérant des sections de E dans celles de E'. Et P^{-1} désigne une parametrix de P. Le résultat ci-dessous est en réalité local, c'est à dire : on peut construire les opérateurs et le fibré du lemme 3.6 au voisinage de tout compact de $\partial\Omega$, mais peut-être pas globalement si $\partial\Omega$ n'est pas compacte.

Lemme 3.6 (i) il existe un fibré vectoriel F sur $\partial\Omega$, un noyau de Green singulier analytique $G' : C_0^\infty(E', \bar\Omega) \to C^\infty(E, \bar\Omega)$ un opérateur trace analytique $T' : C_0^\infty(E', \bar\Omega) \to C^\infty(F, \partial\Omega)$ un noyau de Poisson analytique $K' : C_0^\infty(F, \partial\Omega) \to C^\infty(E, \bar\Omega)$ tels que

$$(3.26) \qquad 1_E \sim P_\Omega\,(P^{-1}_\Omega + G') + K'\, T'$$

(ii) il existe un nombre entier m
un noyau de Green singulier analytique $G'' : C_0^\infty(E', \bar\Omega) \to C^\infty(E, \bar\Omega)$
un noyau de Poisson analytique $K'' : C_0^\infty(E^m, \partial\Omega) \to C^\infty(E, \bar\Omega)$
tels que

$$(3.27) \quad 1_E \sim (P_\Omega^{-1} + G'')\, P_\Omega + K''\, \gamma_m$$

Comme plus haut, le signe $\sim$ signifie que les deux opérateurs ont même symbole complet. En outre, pour abréger la notation, on a omis de tronquer (comme au théorème 2.8). Et nous continurons (incorrectement) à omettre de tronquer ci-dessous.

Preuve du lemme 3.6 :
D'après le théorème 2.8,

$$(3.28) \quad G_1 = 1 - P_\Omega\, P_\Omega^{-1}$$

et

$$(3.29) \quad G_2 = 1 - P_\Omega^{-1}\, P_\Omega$$

sont des noyaux de Green singuliers analytiques, de degré 0.

Le lemme 3.5 fournit alors (localement) un fibré F sur $\partial\Omega$, et des opérateurs G'_1, K'_1, T'_1, G''_1, K''_1, et un nombre m comme dans le lemme ci-dessus, tels que

$$P_\Omega\,(P_\Omega^{-1} + G'_1) + K'_1\, T'_1 = 1 - g'_1$$

$$(P_\Omega^{-1} + G''_1)\, P_\Omega + K''_1\, \gamma_m = 1 - g''_1$$

où g'_1, g''_1 sont des noyaux de Green singuliers analytiques de degré -1. Mais alors, d'après la proposition 2.10, il existe deux autres noyaux de Green singuliers analytiques de degré -1 : g'_2, g''_2 tels que

$$(1 - g'_1)(1 + g'_2) \sim 1_{E'}$$
$$(1 + g''_2)(1 - g''_1) \sim 1_E$$

Il suffit alors de prendre

$$(3.30) \quad G' = G'_1\,(1 + g'_2) + P_\Omega^{-1}\, g'_2 \ ; \quad T' = T'_1\,(1 + g'_2)$$
$$K' = K'_1$$
$$G'' = (1 + g''_2)\, G''_1 + g''_2\, P_\Omega^{-1} \ ; \quad K'' = (1 + g''_2)\, K''_1$$

Nous terminons maintenant la preuve du théorème 3.1 :

Le lemme 3.6 (i) fournit un fibré F sur $\partial\Omega$, et un noyau de Poisson analytique $K : C_0^\infty(F, \partial\Omega) \to C^\infty(E', \bar{\Omega})$ tels que l'opérateur

(3.31) $A = (P_\Omega \oplus K)$

possède une parametrix à droite B', de matrice :

(3.32) $B' = \begin{pmatrix} P_\Omega^{-1} + G' \\ T' \end{pmatrix}$

on a donc

(3.33) $A \cdot B' \sim 1_{E'}$

La deuxième partie du lemme 3.6 nous fournit alors un autre opérateur $B'' : C_0^\infty(E', \bar{\Omega}) \to C^\infty(E, \bar{\Omega}) \oplus C^\infty(F, \partial\Omega)$ de matrice :

(3.34) $B'' = \begin{pmatrix} P_\Omega^{-1} + G'' \\ C \end{pmatrix}$

et un opérateur $K'' : C_0^\infty(E^m \oplus F, \partial\Omega) \to C^\infty(E, \bar{\Omega}) \oplus C^\infty(F, \partial\Omega)$ de matrice :

(3.35) $K'' = \begin{pmatrix} k'' \\ 1_F \end{pmatrix}$

(k'' désigne le noyau de Poisson du lemme 3.6)

tels que

(3.36) $B'' \; A + K'' \; \gamma_m \sim 1_{E \oplus F}$

Si on pose

(3.37) $L' = 1 - B' \; A$

il résulte de (3.33), puis (3.36) qu'on a

(3.38) $A \; L' \sim 0$

$L'^2 \sim L'$

(3.39) $L' \sim K'' \; \gamma_m \; L'$

Posons alors

(3.40) $K^o = L' \; K''$

(3.41) $L = K^o \gamma_m = L' \; K'' \; \gamma_m$

(3.42) $B = (1 - L) \; B'$

On a alors $A\ L \sim O$; $L\ L' \sim L'$ (à cause de (3.38), (3.39)

et par suite $(1-L)\ (1-L') \sim (1-L)$

et finalement

$$(3.43)\quad A\,B = A\,(1-L)\,B' \sim A\,B' \sim 1_{E'}$$

$$(3.44)\quad B\,A = (1-L)\,B'\,A \sim (1-L)\,(1-L') \sim 1-L = 1-K^{o}\gamma_m$$

C'est exactement le résultat annoncé dans le théorème 3.1 . Le reste du théorème en découle immédiatement ((3.43) et (3.44) signifient que moralement , $K^{o}\gamma_m$ est un projecteur sur ker A ; il est clair alors que $H^{o} = \gamma_m K^{o}$ est aussi un "projecteur" ie. vérifie $H^{o2} \sim H^{o}$)

Exemple : Soit Ω le disque complexe $|z| \leq 1$. Soit A l'opérateur de Cauchy-Riemann : $A = \partial/\partial\bar{z} = 1/2\ (\partial/\partial x + i\,\partial/\partial y)$

Désignons par A^{-1} l'opérateur de convolution par la solution élémentaire usuelle de A .

On a alors

$$A\,\tilde{f} = \widetilde{A_{\Omega} f} + z/2\ f_o\,\delta$$

(où δ désigne la mesure usuelle du cercle bord $\partial\Omega$)

Par suite

$$f = A^{-1}\ A\,f + K\,f_o$$

où le noyau de Poisson K est défini par :

$$K(u) = A^{-1}(1/2\ z\,u\,\delta)\,/\Omega = 1/2i\pi \int_{\partial\Omega} u(t)\ dt\,/\,t-z$$

(C'est bien sur l'opérateur qui correspond à la formule de Cauchy)

Alors $H = \gamma K$ est l'opérateur de Hilbert :

$$H\,\Big(\sum_{-\infty}^{+\infty} a_n z^n\Big) = \sum_{0}^{+\infty} a_n z^n$$

Le symbole principal $\sigma_o(H) = h_o(x', \xi')$ est indépendant de ξ', et vaut 1 pour $\xi' > O$, O pour $\xi' < O$.

Le fibré F_o correspondant n'est donc pas de dimension constante , donc n'est certainement pas nul dans le groupe de Grothendieck relatif : On retrouve le fait qu'il n'y a pas de probléme aux limites elliptique associé à $\partial/\partial\bar{z}$.

Bibliographie : on trouvera dans les exposés de R. T. Seeley et de A. P. Calderon une bibliographie concernant les opérateurs pseudo-différentiels en général . Nous limitons ici à ce qui concerne plus spécialement les opérateurs analytiques et les problèmes aux limites elliptiques

[1] L. Boutet de Monvel et P. Krée - Pseudo-differential operators and Gevrey classes - Ann. Inst. Fourier , 1967 , 295 - 323

[2] W. Margulies - Thèse - Brandeis University , 1966

[3] L. Boutet de Monvel - Comportement d'un opérateur pseudo-différentiel sur une variété à bord I et II , Journ. d'Analyse Math. , Jerusalem XVII (1966) 241-304

[4] L. Boutet de Monvel - Opérateurs pseudo-différentiels analytiques et problèmes aux limites elliptiques - à paraître dans Ann. Inst. Fourier

[5] R. T. Seeley - Singular Integrals and Boundary Value Problems - Amer. J. Math. vol 88 , 781-809

[6] M. I. Visik et G. I. Eskin - Equations en convolutions dans un domaine borné , Uspeki Mat. Nauk. , XX , 3 (123) (1965) , 89-152

[7] M. I. Visik et G. I. Eskin - Equations en convolutions dans un domaine borné , Mat. Sbornik , t. 89 (111) , n°1 (1966) , 65-110

CENTRO INTERNAZIONALE MATEMATICO ESTIVO

(C.I.M.E.)

A.P. CALDERON

A PRIORI ESTIMATES FOR SINGULAR INTEGRAL OPERATORS

Corso tenuto a Stresa dal 26 Agosto al 3 Settembre 1968

A PRIORI ESTIMATES FOR SINGULAR INTEGRAL OPERATORS

by

A.P. CALDERÓN
(University of Chicago)

In this paper we develop the theory of singular integral operators with finitely differentiable symbols and apply it to the derivation of a priori inequalities for these and pseudo-differential operators. In section 1 we treat singular integral operators and introduce the pseudo-differential operators as was done before they were given their name, namely as compositions of singular integral operators with powers of the operator Λ. In our opinion this is the correct point of view since in the finitely differentiable case these do not form an algebra, but merely a module over the algebra of singular integral operators. The differentiability assumptions made here are designed to yield self-adjoint algebras which are specially suited for the treatment of the L^2 theory, and are not too far removed from the best possible. Relaxing these conditions substantially, as was done in [1], causes the loss of self-adjointness. In section 2 we discuss the action of our operators on rapidly oscillating functions with small support and obtain, as a byproduct, a new representation for the algebras under consideration which illuminates the negative results on inequalities which are discussed in section. 4.

In section 3 we discuss what we have called almost positive operators. The results here are essentially generalizations of the so called strong Garding inequality and are the foundation on which the a priori inequalities rest. Finally, in section 4 we discuss the inequalities describing the general methods leading to them. These methods are well known (see [2] and [5], for example) but in conjunction with the material of section 3 they yield new and stronger results. For the sake of avoiding cumbersome calculations and formulas we apply these methods only to some simple and well known examples, but we hope that their scope, which extends much farther, will have been made clear. We conclude this section with a brief discussion of the negative results which follow as a consequence of the material in section 2.

Notation and background

Throughout this paper we employ the notation

$$x = (x_1, x_2, \ldots, x_k), \qquad y = (y_1, y_2, \ldots, y_k)$$

$$tx = (tx_1, tx_2, \ldots, tx_k), \quad x+y = (x_1+y_1, x_2+y_2, \ldots, x_k+y_k)$$

$$|x| = (x_1^2+x_2^2+ \ldots +x_k^2)^{1/2}$$

for points of the k-dimensional Euclidean space E^k. Small greek letters will usually stand for multi-indices $\alpha = (\alpha_1, \ldots, \alpha_k)$ of non-negative integers and we will use systematically the following abbreviations

$$|\alpha| = \alpha_1 + \alpha_2 + \ldots + \alpha_k, \quad \alpha! = \alpha_1! \alpha_2! \ldots \alpha_k!$$

$$x^\alpha = x_1^{\alpha_1} x_2^{\alpha_2} \ldots\ldots x_k^{\alpha_k}, \quad \left(\frac{\partial}{\partial x}\right)^\alpha = \left(\frac{\partial}{\partial x_1}\right)^{\alpha_1} \ldots \left(\frac{\partial}{\partial x_k}\right)^{\alpha_k}$$

In order to distinguish differential operators from derivatives we will write $\left(\frac{\partial}{\partial x}\right)^\alpha$ for the operator and $\partial_x^\alpha f$ or f_α for the corresponding derivative of the function f.

The letter c will stand for constants which may be different in different occurrences.

For a real number s, L_s^2 will be the space of distributions f in R^k whose Fourier transforms $\hat{f}$ have the property that $\hat{f}(x)(1+|x|)^s$ is square integrable. The reader will be assumed to be familiar with these spaces. A fairly complete discussion of their properties can be found in [3], for example. A knowledge of the properties of expansions in spherical harmonics will also be assumed. We refer the unfamiliar reader to section 3 of [4] where he will find the backround necessary for the purposes of this paper.

1. Singular integral operators

1.1. For a positive integer m, a singular integral operator of class $\mathcal{S}_m$ is an operator of the form

$$(1) \qquad A f = \sum \int p_j(x,z)\, e^{-2\pi i(x\,z)}\, \hat{f}(z)\, dz + S f$$

where $\hat{f}$ is the Fourier transform of the function f in the class $\mathcal{S}$ of infinitely differentiable rapidly decreasing functions and

i) the functions $p_j(x,z)$ are bounded,

ii) for $|z| > c$, $p_j(x,z)$ coincides for each x with a homogeneous function of z of degree $-d_j$, $0 \leq d_j < d_{j+1} < m$,

iii) $\partial_x^{\alpha}\, \partial_z^{\beta}\, p_j(x,z)$ is a bounded continuous function for $|\alpha| \leq 2m - [d_j]$ and all β,

iv) the operator S maps $\mathcal{S}$ into L^2_m and S, $S(\frac{\partial}{\partial x})^{\alpha}$, $(\frac{\partial}{\partial x})^{\alpha} S$ are bounded with respect to the norm of L^2 for all α with $|\alpha| = m$.

This class of operators will be denoted by $\mathcal{J}_m$.

The name "singular integral operator" stems forn an earlier and alternative description of these operators by means of principal value integrals in f which the reader will find in 1.3 Each of these descriptions has its own particular advantages, the latter being specially suited for the study of continuity properties .

Main properties of singular integral operators

1.2 In many applications the precise nature of the term S in (1) is irrelevant, and in such cases the behaviour of the functions $p_j(x,z)$ near $z=0$ can be disregarded. More precisely

If two operators of the form (1) are such that the corresponding functions

$p_j(x,z)$ coincide for $|z|>c$, then their difference belongs to $\mathcal{J}_m$.

Let A be the difference of these two operators. We will show first that A is bounded in L^2. Since A has also the form (1), we can write $A = \sum A_j + S$, where

$$(2)\qquad \begin{aligned} A_j f &= \int p_j(x,z)\, e^{-2\pi i(x\cdot z)}\hat{f}(z)\, dz = \\ &= \int f(x+y) k_j(y,x)\, dy \end{aligned}$$

where $k(y,x)$ is the Fourier transform of $p_j(x,z)$ with respect to z. Now since $p_j(x,z)$ vanishes for $|z|> c$, it is integrable with respect to z uniformly in x, and $k_j(y,x)$ is bounded. Furthermore, since $\partial_z^\beta p_j(x,z)$ is also uniformly integrable with respect to z for all β, it follows that $k_j(y,x)y^\beta$ is bounded for all β. Thus we have

$$|k_j(y,x)| \leq \frac{c}{1+|y|^{k+1}}$$

and the integral in (2) is dominated by the convolution of $|f|$ with an integrable function, which implies that A_j is bounded in L^2. Now, given the differentiability properties of the functions $p_j(x,z)$, the same conclusion holds for $(\frac{\partial}{\partial x})^\alpha A_j$ and $A_j(\frac{\partial}{\partial x})^\alpha$ for all α, $m = |\alpha|$, these operators being of the same form as A_j, or else sums of operators like A_j. Thus A_j belongs to $\mathcal{J}_m$ and desired conclusion follows.

1.3 <u>Operators in $\mathcal{J}_m$ can be extended to bounded operators from L_s to L_{s+m}, $0 \geq s \geq -m$, and conversely, every operator with this property belongs to $\mathcal{J}_m$. Let $\mathcal{J}_0$ denote the class of bounded operators in L^2. Then if A_j is the operator defined by the j-th term of the sum in the righthand side of (1), A_j belongs to $\mathcal{J}_n$ for all n with $n \leq d_j$. Furthermore, A_j can be extended to a bounded operator</u>

<u>from L^2_s to L^2_{s+t} provided that $-m \leq s \leq s+t \leq m$ and $t \leq d_j$. Consequently $\mathcal{J}_m \subset \mathcal{J}_n$ for $m \geq n$, and every operator in $\mathcal{J}_m$ can be extended to a bounded operator in L^2_s for $-m \leq s \leq m$.</u>

Let us prove this. Let S be an operator in $\mathcal{J}_m$. The condition that S and $\left(\frac{\partial}{\partial x}\right)^{\alpha} S$ be bounded in L^2 for $|\alpha| = m$ implies that S can be extended to a continuous operator from L^2 to L^2_m (this is an immediate consequence of the definition of the norm in L^2_m). Similarly, the condition that S and $S\left(\frac{\partial}{\partial x}\right)^{\alpha}$ be bounded in L^2 for $|\alpha| = m$ implies that S can be extended to a bounded operator from L^2_{-m} to L^2. Now by interpolation (see [3], theorem 10), it follows that S can be extended to a bounded operator from L^2_s to L^2_{s+m} for $0 \geq s \geq -m$. The converse of this follows from the fact that $\left(\frac{\partial}{\partial x}\right)^{\alpha}$, $|\alpha| = m$, maps L^2_s continuously into L^2_{s-m}.

Let us consider now the operator A_j. We will assume that $p_j(x, z)$ has the form

$$p_j(x, z) = q(x, z)\, \psi(z)\, |z|^{-d_j}$$

where $\psi(z)$ vanishes near $z=0$ and equals 1 outside a compact set, and $q(x, z)$ is homogeneous of degree 0 in z. Given the differentiability properties of q, which follow from the fact that q and $p|z|^{-d_j}$ coincide for $|z|$ large, we can expand it in a series of spherical harmonics in z

$$q(x, z) = \sum a_{n\ell}(x)\, Y_{n\ell}(z)$$

where for each n, $Y_{n\ell}$ is a complete set of normalized spherical hermonics of degree n, and

(3) $$\left|\partial_x^{\beta}\, a_{n\ell}(x)\right| \leq c\, n^{-r} M$$

where M is a bound for $\left|\partial_x^{\beta}\, \partial_z^{\gamma}\, q(x, z)\right|$, $|\gamma| \leq r$, in $|z| \geq 1$, and c depends on r. Let us set now

$$T_{n\ell} f = \int Y_{n\ell}(z)\, \psi(z)\, |z|^{-d_j} e^{-2\pi i(x.z)} \hat{f}(z)\, dz$$

(4) $$A_{n\ell} f = a_{n\ell}(x)\, f(x)$$

Since $|Y_{n\ell}(z)| \leq c\, n^{1/2(k-2)}$, it follows that the norm of $T_{n\ell}$ as an operator from L^2_s to L_{s+t}, $t \leq d_j$, is dominated by $c\, n^{1/2(k-2)}$. On the other hand, on account of (3), the norm of $A_{n\ell}$ as an operator on L^2_s, $|s| \leq m$, is dominated by $c\, n^{-r} M$. For s a positive integer this is clear; for s a negative integer this follows by duality, and for general s by interpolation. Consequently, the norm of $A_{n\ell} T_{n\ell}$ as an operator from L^2_s to L^2_{s+t}, $-m \leq s \leq s+t \leq m$, $t \leq d_j$, is dominated by $c\, M\, n^{-r+(k-2)/2}$, and since there are at most $c\, n^{k-2}$ harmonics of degree n, and r can be taken arbitrarily large, the series $\sum A_{n\ell} T_{n\ell}$ converges absolutely with respect to this norm to A_j, which has the continuity properties we claimed. From this since L^2_s is continuously embedded in L^2_t for $s \geq t$, our remaining assertions follow readily.

1.4. <u>Operators in $\mathcal{J}_m$ can be extended to bounded operators in L^2 and their extensions form a selfadjoint algebra under composition. Let $\tilde{p}_j(x, z)$ be homogeneous in z and coincide with the function $p_j(x, z)$ in (1) for z sufficiently large. Then the function $\sum \tilde{p}_j(x, z)$ is uniquely determined by the operator A and is called the symbol of A and denoted by $\sigma(A)$. Conversely, $\sigma(A)$ determines A modulo $\mathcal{J}_m$. For the symbols of products and adjoints we have</u>

(5) $$\sigma(AB) = \sum \frac{1}{\alpha!} \left(\frac{i}{2\pi}\right)^{|\alpha|} \left[\partial_z \sigma(A)\right] \left[\partial_x \sigma(B)\right]$$

(6) $$\sigma(A^*) = \sum \frac{1}{\alpha!} \left(\frac{i}{2\pi}\right)^{|\alpha|} \partial_z^\alpha \partial_x^\alpha \overline{\sigma(A)}$$

where the sums are extended over all terms whose degree of homogeneity with respect to z is larger than $-m$. We will denote the the righthand sides of (5) and (6) by $\sigma(A) \circ \sigma(B)$ and $\sigma(A)^{\#}$ respectively.

That operators in $\mathcal{J}_m$ can be extended to bounded operators in L^2 was shown in 1. 3. That the symbol $\sigma(A)$ is uniquely determined by A will be shown in section 2 .

Let us begin then by discussing the formalism of the symbols. We have introduced the operations $p \circ q$ and $p^{\#}$ on the symbols p and q . We want to show that these operations satisfy the formal algebraic laws, namely the associative law for the product $p \circ q$ (the distributive law is clearly satisfied), and that the operation $p \longrightarrow p^{\#}$ is an involution and has the property that $(p \circ q)^{\#} = q^{\#} \circ p^{\#}$. Doing this by straight forward calculation requires proving some non-obvious identities between factorials. We shall therefore use a different approach which has the advantage of illuminating this formalism from an algebraic point of view.

Let us consider differential operators of the form

$$\mathcal{A} = \sum a_{\alpha}(x, z) \left(\frac{\partial}{\partial z}\right)^{\alpha} \tag{7}$$

where the $a_{\alpha}(x, z)$ are sums of infinitely differentiable homogeneous functions of non-positive degree in z. The weight of a monomial operator $a(x, z)\left(\frac{\partial}{\partial z}\right)^{\alpha}$ with $a(x, z)$ homogeneous in z will be, by definition, the order $|\alpha|$ minus the degree of homogeneity of $a(x, z)$. Given two such monomial differential operators $\mathcal{A}_1$ and $\mathcal{A}_2$ of weights w_1 and w_2 , $\mathcal{A}_1 \mathcal{A}_2$ and $\mathcal{A}_1^{\#}$ are readily seen to be sums of operators of weights $w_1 + w_2$ and w_1 respectively. Thus the operators $\mathcal{A}$ form a selfadjoint algebra $\mathcal{A}_{\zeta}$ and given a positive real number s , operators $\mathcal{A}$ which are expressible as a sum of terms of weight larger than or equal to s form a two-sided selfadjoint ideal $\mathcal{J}_s$ in $\mathcal{A}$. In the quotient algebra $\mathcal{A}_s = \mathcal{A}/\mathcal{J}_s$ we have a conjugation induced by the conjugation $\lambda \to \lambda^*$. Let us consider now the following derivations of

(8) $$\mathcal{D}_j \lambda = \frac{\partial \lambda}{\partial x_j} + 2\pi i\,(\lambda z_j - z_j \lambda)$$

where z_j here stands for the operator multiplication by z_j. Evidently $(\mathcal{D}_j \lambda)^* = \mathcal{D}_j \lambda^*$, and $\mathcal{D}_j$ maps $\mathcal{A}_s$ into $\mathcal{A}_{s-1}$ and therefore induces a derivation $\mathcal{D}_j$ of $\mathcal{A}_s$ into $\mathcal{A}_{s-1}$. The set of elements λ of $\mathcal{A}_s$ such that $\mathcal{D}_j \lambda = 0$, $j = 1, 2, \ldots, k$, is therefore a selfadjoint subalgebra $\bar{\mathcal{A}}_s$ of $\mathcal{A}_s$. The elements of $\mathcal{A}$ which reduced modulo $\mathcal{J}_s$ yield elements of $\bar{\mathcal{A}}_s$ satisfy the relations

$$\mathcal{D}_j \lambda \equiv 0 \pmod{\mathcal{J}_{s-1}}$$

which on account of the fact that

$$\mathcal{D}_j \left[\left(\frac{\partial}{\partial z_\ell}\right)^n \right] = 2\pi i\, n \left(\frac{\partial}{\partial z_\ell}\right)^{n-1} \delta_{j\ell}$$

are readily seen to imply that

$$\lambda \equiv \sum \frac{1}{\alpha!}\left(\frac{i}{2\pi}\right)^{|\alpha|} \partial_x^\alpha\, a \left(\frac{\partial}{\partial z}\right)^\alpha \pmod{\mathcal{J}_s}$$

where a is the term of order 0 in λ reduced modulo homogeneous terms of degree less than or equal to $-s$. This function we shall call the symbol $\sigma(\lambda)$ of λ. Now, if λ_1 and λ_2 reduced modulo $\mathcal{J}_s$ yield elements of $\bar{\mathcal{A}}_s$, the same is true for $\lambda_1 \lambda_2$ and λ_1^* and, therefore, $\sigma(\lambda_1 \lambda_2)$ and $\sigma(\lambda_1^*)$ are well defined, and using the formula for differentiation of products one verifies readily that

$$\sigma(\lambda_1 \lambda_2) = \sum \frac{1}{\alpha!}\left(\frac{i}{2\pi}\right)^{|\alpha|} \left[\partial_z^\alpha \sigma(\lambda_2)\right]\left[\partial_x^\alpha \sigma(\lambda_1)\right]$$

$$\sigma(\lambda_1^*) = \sum \frac{1}{\alpha!}\left(\frac{i}{2\pi}\right)^{|\alpha|} \partial_z^\alpha\, \partial_x^\alpha\, \overline{\sigma}(\lambda_1)$$

where the sums are extended over all terms of degree of homogeneity larger than -s. Setting s=m, and comparing with (5) and (6) we find that

$$\sigma(\lambda_1 \lambda_2) = \sigma(\lambda_2) \circ \sigma(\lambda_1)$$
$$\sigma(\lambda_1^*) = \sigma(\lambda_1)^*$$

from which the formal properties of the operations on symbols defined by (5) and (6) follow at once, at least for infinitely differentiable symbols. However, these formal laws are merely algebraic identities which, if valid for infinitely differentiable functions, must hold in the more general case under consideration, as is readily seen by a passage to the limit.

Let us turn now to the singular integral operators. We will begin by showing that if A is in $\mathcal{S}_m$ so is A^*. If A is in $\mathcal{J}_m$ this is clear. Thus it will be enough to show that the adjoints of operators like the A_j in 1.2 are in $\mathcal{S}_m$. This we will do in the case when p has the form

$$p(x,z) = a(x)\; Y(z)\; \psi(z)\; |z|^{-d}$$

where

$$\psi(z) = |z|^d \int_0^1 \varphi(tz)\; t^d \frac{dt}{t}$$

with φ a spherically symmetric function in C^∞ with support in $\frac{1}{2} \leq |z| \leq 1$, and $Y(z)$ an infinitely differentiable homogeneous function of degree 0. The general case will then follow by addition. Since $Y(z)$ is homogeneous of degree 0, we have

$$\begin{aligned} p(x,z) &= a(x)\, Y(z) \int_0^1 \varphi(tz)\, t^d \frac{dt}{t} \\ &= a(x) \int_0^1 Y(tz)\, \varphi(tz)\, t^d \frac{dt}{t} \\ &= a(x) \int_0^1 g(tz)\, t^d \frac{dt}{t} \end{aligned}$$

where $g(z) = Y(z)\varphi(z)$. Thus the operator under consideration has the form

$$A\ f = a(x) \int e^{-2\pi i(x.z)} \hat{f}(z) \int_0^1 g(tz)\, t^d \frac{dt}{t}\, dz$$

$$= a(x) \int_0^1 t^d \frac{dt}{t} \int e^{-2\pi i(x.z)} g(tz)\, \hat{f}(z)\, dz$$

and applying Plancherel's formula to the inner integral we find

(9)
$$A\ f = a(x) \int_0^1 t^d \frac{dt}{t} \int h\left(\frac{y-x}{t}\right) t^{-k} f(y)\, dy$$

where $h = \hat{g}$. Now, if $d>0$ this integral is absolutely convergent and interchanging the order of integration and setting

(10)
$$k(x) = \int_0^1 h\left(\frac{x}{t}\right) t^{-k+d} \frac{dt}{t}$$

we obtain

(11)
$$A\ f = a(x) \int k(x-y)\, f(y)\, dy$$

If $d=0$ this representation is still valid if the integral is interpreted in the sense of principal values provided that the mean value of $Y(z)$ on the unit sphere $|z| = 1$ vanishes. In fact, if this is the case then $g(z) = Y(z)\varphi(z)$ is orthogonal to every spherically symmetric function and the same is true of $h(\frac{x}{t})$ for all $t>0$. Thus in (9) we can replace $f(y)$ by $f(y) - f(x)\eta(x-y)$ where $\eta(x)$ is spherically symmetric, in C_0^∞ and $\eta(0)=1$. After this substitution (9) is again absolutely convergent and instead of (11) we obtain

$$A\ f = a(x) \int k(x-y) \left[f(y) - f(x)\, \eta(x-y) \right] dy$$

where the integral is also absolutely convergent. Now, since $h(\frac{x}{t})$ is orthogonal to every spherically symmetric function, $k(x)$ is orthogonal to every such function vanishing near the origin. Thus for every ε', $\varepsilon > 0$

$$\int_{|x-y|>\varepsilon} k(x-y)\ f(x)\ \eta\ (x-y)\ dy = 0$$

and consequently

$$\lim_{\varepsilon\to 0} \int_{|x-y|>\varepsilon} k(x-y)\ f(y)\ dy = \lim_{\varepsilon\to 0} \int_{|x-y|>\varepsilon} k(x-y)\left[f(y)-f(x)\eta(x-y)\right]dy = A\ f$$

If, on the other hand, $d=0$ and $Y(z)$ has mean value M on the unit sphere $|z| = 1$, we set $Y(z) = M + Y_1(z)$. The function $Y_1(z)$ has mean value 0 on $|z| = 1$ and gives rise to an operator which has a representation like in (11). The term M gives rise to the operator

$$A\ f = a(x) \int M\psi(z)\ \ e^{-2\pi i(x\cdot z)}\ \hat{f}(z)\ dz\ .$$

Assuming, as we may, that $\psi(z) = 1$ for $|z| \geq 1$, this operator differs from

$$A_1 f = a(x) \int M\ e^{-2\pi i(x\cdot z)}\hat{f}(z)\ dz = a(x)\ M\ f(x)$$

by an element of $\mathcal{J}_m$. Since A_1^* obviously belongs to $\mathcal{J}_m$, the same is true of A^* and clearly (6) holds for $\sigma(A^*)$.

Consequently we only have to consider operators which have the representation (11). We note that the function $k(x)$ there is the inverse Fourier transform of $Y(z)\psi(z)|z|^{-d}$. Now, from this expression of A we find that

$$(12) \qquad A^*\ f = \int \overline{k}(y-x)\ a(y)\ \ dy$$

where the integral is to be interpreted as a principal value integral if necessary, and expanding $\overline{a}(y)$ by Taylor's formula

$$(13) \qquad A^* f = \sum \frac{1}{\alpha!}\ \overline{a}_\alpha(x) \int \overline{k}(y-x)\ (y-x)^\alpha f(y)dy + \int R(y,x)\overline{k}(y-x)\ f(y)\ dy$$

Here the sum is taken over all α with $|\alpha| < 2m - [d]$. Since $k(x)\, x^{\alpha}$ is the inverse Fourier transform of $(\frac{1}{2\pi i})\partial_z^{\alpha}\left[Y(z)\psi(z)\,|z|^{-d}\right]$ $\bar{k}(-x)\,(-x)^{\alpha}$ is the inverse Fourier transform of $(\frac{i}{2\pi})^{|\alpha|}\partial_z^{\alpha}\left[Y(z)\psi(z)\,|z|^{-d}\right]$ (we assume $Y(z)$ to be real valued). Consequently the terms of the sum above can be expressed as

$$\sum \int \frac{1}{\alpha!}\left(\frac{i}{2\pi}\right)^{|\alpha|} e^{-2\pi i(x\cdot z)}\left[\partial_x^{\alpha}\partial_z^{\alpha}\bar{p}(x,z)\right]\hat{f}(z)\,dz \quad .$$

The sum of terms with $|\alpha| + d < m$ here is an operator A_1 in $\mathcal{S}_m$ with symbol given by (6) . Consequently we have

$$A^{*}f = A_1 f + \sum \frac{1}{\alpha!}\left(\frac{i}{2\pi}\right)^{|\alpha|}\int e^{-2\pi i(x\cdot z)}\left[\partial_x^{\alpha}\partial_z^{\alpha}\bar{p}(x,z)\right]\hat{f}(z)\,dz + \tag{14}$$

$$+\int R(y,x)\,\bar{k}(y-x)\,f(y)\,dy$$

where the sum is extended over the remaining α, that is all α with $|\alpha| + d \geq m$. We will show now that this sum, which we will denote by S , as well as the last term in (14) , remain bounded in L^2 after composition on the right with differentiation of order m . For S this is immediate, given the special form of $p(x,z)$. Furthermore, we see that the norm of $S(\frac{\partial}{\partial x})^{\alpha}$, $|\alpha| = m$, does not exceed $c\,N\,M$ where N and M are bounds for $a(x)$ and $Y(z)$ and their derivatives up to order $2m - [d]$ and $2m - [d]+k$ respectively, and c depends on ψ , m and d.

To estimate the last term in (14) we need estimates of $R(y,x)$ $k(x)$ and some of their derivatives. Let now M and N be as above and let $r = 2m - [d]$. Then we have

$$\left|\partial_y^{\beta} R(y,x)\right| \leq c\,N\,|x-y|^{r-|\beta|}(1+|x-y|)^{-1},\ |\beta| \leq r \tag{15}$$

On the other hand, for $h = \hat{g}$, $g(z) = Y(z)\varphi(z)$ we have

$$\left|\partial_x^\beta h(x)\right| \leq \frac{cM}{(1+|x|)^{r+k}}$$

From this inequaltiy, differentiating under the integral sign in (10), substituting and setting $|x| = ts$ we obtain

$$\left|\partial_x^\beta k(x)\right| \leq |x|^{-k+d-|\beta|} \int_{|x|}^{\infty} \frac{c\,M}{(1+s)^{r+k}}\, s^{k-d+|\beta|} \frac{ds}{s}$$

and

$$\left|\partial_x^\beta k(x)\right| \leq c\,M \frac{|x|^{-k+d-|\beta|}}{(1+|x|)^{r+d-m}}, \quad |\beta| \leq m,$$

Turning to (14) we let now S be the operator represented by the last term. Then

$$S\,\partial_x^\alpha f = \int R(y,x)\,\bar{k}(y-x)\,\partial_y^\alpha f(y)\,dy$$

and integrating by parts

$$S\,\partial_x^\alpha f = \sum \int c_{\beta\gamma} \left[\partial_y^\beta R(y,x)\right]\left[\partial_y^\gamma \bar{k}(y-x)\right] f(y)\,dy$$

where the $c_{\beta\gamma}$ are constants and $\beta+\gamma = \alpha$. Now if $|\alpha| = m$ using the estimates for $R(y,x)$ and $k(x)$ obtained above we find that the integrals in the preceding sum are dominated by

$$c\,N\,M \int \frac{|x-y|^{-k-m+r+d}}{(1+|x-y|)^{r+d-m+1}}\, f(y)\,dy$$

and by Young's theorem on convolutions we find that $S\left(\frac{\partial}{\partial x}\right)^\alpha$, $|\alpha| = m$, is bounded in L^2 and its norm is dominated by $c\,N\,M$.

Summarizing we have obtained the following result: for an operator A_j with

$$p(x,z) = a(x)\,Y(z)\,\mathcal{Y}(z)\,|z|^{-d}$$

we have $A_j = A' + S$, where A' is a sum of operators like the A_j themselves, $\sigma(A') = \sigma(A_j)^{\#}$, and $S\left(\frac{\partial}{\partial x}\right)^\alpha$, $|\alpha| = m$, is bounded in L^2 and has norm dominated by $c\,N\,M$, where N and M are bounds for $a(x)$ and $Y(z)$ and their derivatives up to order $2m-[d]$ and $2m-[d]+k$ respectively.

Now we generalize this result replacing $a(x)\,Y(z)$ by a function

$q(x,z)$, homogeneous of degree 0 in z and satisfying the differentiability conditions iii) in 1.1. For this purpose we expand $q(x,z)$ in spherical harmonics as in 1.3 and obtain

$$A_j = \sum A_{n\ell} T_{n\ell} \tag{13}$$

where the operators on the right are defined by (4). Applying the preceding result to each of the terms on the right we have

$$A_j^* = \sum C_{n\ell} + S_{n\ell}$$

where $\sigma(C_{n\ell}) = [a_{n\ell}(x)\, Y_{n\ell}(z)]^{\#}$. Since for the derivatives of $Y_{n\ell}$ we have the estimate

$$\left| \partial_z^{\beta} Y_{n\ell}(z) \right| \leq c\, |z|^{-|\beta|} n^{1/2(k-2)+|\beta|}$$

and the $a_{n\ell}(x)$ satisfy the inequalities (3), the series $\sum C_{n\ell}$ converges to an operator C with $\sigma(C) = \sigma(A_j)^{\#}$. On the other hand, the norm in L^2 of $S_{n\ell}(\frac{\partial}{\partial x})^{\alpha}$, $|\alpha| = m$, can be estimated by using the preceding inequalities for $a_{n\ell}(x), Y_{n\ell}(z)$, and turns out to be dominated by $c\, M\, n^{2(k+m)-1-r}$ where M and r are now as in (3). Since for each n there are at most $c\, n^{k-2}$ terms $S_{n\ell}$, taking $r = 3k+2m-1$, we conclude that

$$A_j^* = C+S \tag{14}$$

where $\sigma(C) = \sigma(A_j)^{\#}$ and $S(\frac{\partial}{\partial x})^{\alpha}$, $|\alpha| = m$, has norm of the order $c\,M$ in L^2 where M is a bound for $\left| \partial_x^{\beta} \partial_z^{\gamma} q(x,z) \right|$, $|\beta| \leq 2m-[d]$, $|\gamma| \leq 3k+2m-1$. On account of what was seen in 1.2 this is also valid, except for the estimate of the norm of $S(\frac{\partial}{\partial x})^{\alpha}$, even if the function $p(x,z)$ does not have the particular form postulated so far. By addition we find now that if A is in $\mathcal{S}_m$ then $A^* = C+S$, where C is likewise n $\mathcal{S}_m$, $\sigma(C) = \sigma(A)^{\#}$, and $S(\frac{\partial}{\partial x})^{\alpha}$, $|\alpha| = m$, is bounded in L^2. Thus, in order to show that A^* is in $\mathcal{S}_m$ and that its symbol can be calculated by means of (6), we have to show that S is in $\mathcal{J}_m$. That S is bounded in L^2 follows from the boundedness of A^* and C. That $S(\frac{\partial}{\partial x})^{\alpha}$, $|\alpha| = m$, is also bounded in L^2, we have already shown. Consequently there remains to show that $(\frac{\partial}{\partial x})^{\alpha} S$ or, which amounts to the same, $S^*(\frac{\partial}{\partial x})^{\alpha}$ is bounded in L^2. For this purpose we take adjoints in $A^* = C+S$ and apply the preceding result to C^*. Thus we obtain $C^* = C_1+S_1$ where $\sigma(C_1) = \sigma(C)^{\#} = \sigma(A)^{\#\#} = \sigma(A)$, and $S^* = A-C_1-S_1$. Since $\sigma(C_1) = \sigma(A)$, $A-C_1$ belongs to $\mathcal{J}_m$ and we find that

$$S^*(\frac{\partial}{\partial x})^{\alpha} = (A - C_1 - S_1)(\frac{\partial}{\partial x})^{\alpha}, \qquad |\alpha| = m$$

is bounded in L^2 as we wished to show.

Let us turn now to the composition of singular integral operators. First let us observe that the operators A_j defined by the various terms of the sum in (1) and the operator S there retain their continuity properties described in 1.3 after being continuously extended to L^2. Thus it is clear that composition of operators in $\mathcal{S}_m$ with operators in $\mathcal{J}_m$ yield operators in $\mathcal{S}_m$. Consequently it will suffice to consider products of the form $A B^*$ where A and B have the form

$$A f = \int p(x,z)\,\psi(z)\,|z|^{-d}\, e^{-2\pi i(x\cdot z)}\hat{f}(z)\, dz$$

$$B f = \int q(x,z)\,\eta(z)\,|z|^{-d'}\, e^{-2\pi i(x\cdot z)}\hat{f}(z)\, dz$$

where p and q are homogeneous of degree 0 in z and satisfy the differentiability conditions iii) in $|z| \geq 1$, ψ and η are infinitely differentiable vanish near the origin and have the properties $\psi(z)\eta(z) = \eta(z)$, $\eta(z) = 1$ for $|z| \geq 1$, and

$$\eta(z) = |z|^{d+d'} \int_0^1 \varphi(tz)\, t^{d+d'} \frac{dt}{t}$$

where $\varphi(z)$ is infinitely differentiable, spherically symmetric and has support in $\frac{1}{2} \leq |z| \leq 1$.

If $d+d' \geq m$, then, on account of the continuity properties of A and B described in 1.3, it follows that $A B^*$ is in $\mathcal{J}_m$. If $d+d'<m$, we expand $p(x, z)$ in spherical harmonics and obtain

$$A = \sum A_{n\ell} T_{n\ell}$$

the operators on the right here being defined by (4). If we assume, as we may, that $\psi(z)$ and $Y_{n\ell}(z)$ are real, then $T_{n\ell}$ is selfadjoint and $T_{n\ell} B^* = (B\, T_{n\ell})^*$. Now, we have

$$B\, T_{n\ell} f = \int q(x, z)\, \eta(z)\, |z|^{-d-d'}\, Y_{n\ell}(z)\, e^{-2\pi i(x \cdot z)} \hat{f}(z)\, dz$$

and we can apply to this operator our results on adjoints. Thus we find that $(B\, T_{n\ell})^* = C_{n\ell} + S_{n\ell}$, where $\sigma(C_{n\ell}) = [\sigma(B)\, Y_n(z)\, |z|^{-d}]^{\#}$ and $S_{n\ell}\left(\frac{\partial}{\partial x}\right)^{\alpha}$, $|\alpha| = m$, is bounded in L^2 with norm of the order $n^{4k+2m-1}$ Consequently

$$A = \sum A_{n\ell} C_{n\ell} + A_{n\ell} S_{n\ell}$$

Now, given the differentiability properties of $a_{n\ell}$ we see that $A_{n\ell} C_{n\ell}$ is an operator in $\mathcal{J}_m$ and

$$\begin{aligned}
\sigma(A_{n\ell} C_{n\ell}) &= \sigma(A_{n\ell}) \circ \sigma(C_{n\ell}) \\
&= \sigma(A_{n\ell}) \circ [\sigma(B) \circ \sigma(T_{n\ell})]^{\#} \\
&= \sigma(A_{n\ell}) \circ \sigma(T_{n\ell}) \circ \sigma(B)^{\#} \\
&= \sigma(A_{n\ell} T_{n\ell}) \circ \sigma(B)^{\#}
\end{aligned}$$

On the other hand, given the estimates on the derivatives of the $Y_{n\ell}$ and (3) it is clear that the series $\sum A_{n\ell} C_{n\ell}$ converges to an operator C in $\mathcal{J}_m$ whose symbol is $\sigma(A) \circ \sigma(B)^{\#}$, and the series $\sum A_{n\ell} S_{n\ell}$ converges to an operator S such that $S(\frac{\partial}{\partial x})^{\alpha}$, $|\alpha| = m$, is bounded in L^2. Thus we have

$$A \cdot B^* = C + S$$

where C is in $\mathcal{J}_m$ and $\sigma(C) = \sigma(A) \circ \sigma(B)^{\#}$. Thus, if we show that S and $S^*(\frac{\partial}{\partial x})^{\alpha}$ or, which amounts to the same, $(\frac{\partial}{\partial x})^{\alpha} S$ are bounded in L^2 for $|\alpha| = m$, it will follow that S is in $\mathcal{J}_m$, that $A B^*$ is in $\mathcal{J}_m$ and that its symbol is given by (5). That S is bounded in L^2 follows from the boundedness of A, B and C. On the other hand, applying the preceding result to $B A^*$ we obtain

$$B A^* = C_1 + S_1 = (A B^*)^* = C^* + S^*$$

$$\sigma(C_1) = \sigma(B) \circ \sigma(A)^{\#} = [\sigma(A) \circ \sigma(B)^{\#}]^{\#} = \sigma(C^*)$$

Since the operator S_1 has the property that $S_1(\frac{\partial}{\partial x})^{\alpha}$, $|\alpha| = m$, is bounded in L^2, and since $\sigma(C_1) = \sigma(C^*)$, it follows that $C_1 - C^*$ is in $\mathcal{J}_m$ and that

$$S^*(\frac{\partial}{\partial x})^{\alpha} = (C_1 - C^* + S_1)(\frac{\partial}{\partial x})^{\alpha}, \quad |\alpha| = m$$

is bounded in L^2, as we wished to show

1.5. Pseudo-differential operators. We define these operators by means of singular integral operators and the operator Λ. Let $\varphi(z)$ be real infinitely differentiable and such that $\varphi(z) \geq 1$, $\varphi(z) = |z|$ for $|z| \geq 2$. Then for s real we define Λ^s by $(\Lambda^s f)^{\wedge} = \varphi(z)^s \hat{f}(z)$. The main properties of Λ^s are immediate consequences of its definition. First of all, $\Lambda^{s_1+s_2} = \Lambda^{s_1} \Lambda^{s_2}$. Then, Λ^s maps L^2_{r+s} continuously onto L^2_r, and finally, for $s < 0$ the operator Λ^s belongs to all $\mathcal{J}_m$.

A right (left) pseudo-differential operator of class m and order s, $\infty > s > -\infty$, is an operator of the form $P = \Lambda^s A$ $(P = A\Lambda^s)$ where A is in $\mathcal{S}_m$. Evidently these classes are closed under composition on the right (left) with operators in $\mathcal{S}_m$, that is, they are a right (left) module over $\mathcal{S}_m$. A right pseudo-differential operator can be composed freely on the right with a left pseudo-differential operator giving rise to what we call mixed pseudo-differential operators. The continuity properties of all these operators follow readily from those of Λ^s and of the singular integral operators.

Between these classes of pseudo-differential operators there are certain inclusion relations which are consequences of the following fact:

<u>Let $m-1 \geq |s|$. Then if A is in $\mathcal{S}_m$</u>

$$\Lambda^{s_1} A \Lambda^{s+s_2} = \Lambda^{s_1+s} B \Lambda^{s_2}$$

<u>where B is an operator in $\mathcal{S}_n$ with $m \geq n \geq [m - |s|]$.</u>

<u>Furthermore</u>

$$\sigma(B) = \sum \frac{1}{\alpha!}\left(\frac{i}{2\pi}\right)^{|\alpha|}\left[\partial_z^\alpha |z|^{-s}\right]\left[\partial_x^\alpha \sigma(A)\, |z|^{s}\right] \tag{15}$$

<u>where the sum is extended over all terms whose degree of homogeneity in z is larger than $-n$.</u>

Let us show this. Suppose first that s is positive. Then Λ^{-s} and $C = \Lambda^{-s} A$ are operators in $\mathcal{S}_m$. Furthermore C has the form

$$C f = \sum \int p_j(x, z)\, e^{-2\pi i(x\cdot z)}\, \hat{f}(z)\, dz + S f$$

where the $p_j(x, z)$ coincide with homogeneous functions of z for $|z| \geq c$ and the corresponding degrees of homogeneity are less than or equal to $-s$. Consequently

$$B\, f = C\Lambda^s f = \sum \int p_j(x,z)\, \varphi(z)^s \; e^{-2\pi i(x\cdot z)}\hat{f}(z)\, dz + S\Lambda^s f$$

still has the form (1) . Let now $q_j(x,z) = p_j(x,z)\,\varphi(z)^s$. Since for $|z| \geq c$ the function $p_j(x,z)$ is homogeneous of degree $-d_j$ in z , $q_j(x,z)$ is homogeneous of degree $-d_j+s$ for $|z| \geq c$. Let now n be the largest integer with the property that $\partial_x^\alpha \partial_z^\beta q_j(x,z)$ is bounded for $|\alpha| \leq 2n - [d_j - s]$ and the operators $S\Lambda^s$ and

$$B_j f = \int q_j(x,z)\, e^{-2\pi i(x\cdot z)}\hat{f}(z)\, dz\,, \qquad d_j - s \geq n$$

belong to $\mathcal{J}_n$. As readily seen, on account of the differentiability properties of the $p_j(x,z)$ and of 1.3 the integer $[m-s]$ has this property and consequently $n \geq [m-s]$. But then $B = C\Lambda^s = \Lambda^{-s} A\Lambda^s$ belongs to $\mathcal{J}_n$ and $\Lambda^{s_1} A\Lambda^{s+s_2} = \Lambda^{s_1+s} B\Lambda^{s_2}$. In order to calculate the symbol of B we observe that $\sigma(B)$ is obtained by deleting all terms of degree larger than $-n$ in the expression for $\sigma(C)\,|z|^s$. Therefore, on account of (5) , (15) follows .

Let now s be negative and let

$$B = \Lambda^{-s} A\Lambda^s = (\Lambda^s A^* \Lambda^{-s})^* .$$

Since A^* is in $\mathcal{J}_m$, by the preceding result it follows that $\Lambda^s A^* \Lambda^{-s} = B^*$ and B are in $\mathcal{J}_n$ with $n \geq [m - |s|]$. To calculate the symbol of B we observe that by the preceding result , since $B\Lambda^{-s} = \Lambda^{-s} A$, the symbol of A as an element of $\mathcal{J}_n$ is given by

$$\sigma(A) = \sum \frac{1}{\alpha!}\left(\frac{i}{2\pi}\right)^{|\alpha|} \left[\partial_z^\alpha |z|^s\right]\left[\partial_x^\alpha \sigma(B)\, |z|^{-s}\right]$$

where the sum is extended over all terms of degree larger than -n in z . Deleting on both sides of this equation all terms of degree less than or equal to $-n+|\beta|$ and differentiat ng we obtain

$$\frac{1}{\beta!}\left(\frac{i}{2\pi}\right)^{|\beta|}\left[\partial_z^{\beta}|z|^{-s}\right]\left[\partial_x^{\beta}\sigma(A)\,|z|^{s}\right] =$$

$$=\sum\frac{1}{\beta!\alpha!}\left(\frac{i}{2\pi}\right)^{|\alpha+\beta|}\left[\partial_z^{\beta}|z|^{-s}\right]\left[\partial_z^{\alpha}|z|^{s}\right]\left[\partial_x^{\alpha+\beta}\sigma(B)\right]$$

Summing over all β and observing that for $|\gamma|>0$

$$0=\partial_z^{\gamma}1=\partial_z^{\gamma}(|z|^{-s}|z|^{s})=\sum_{\alpha+\beta=\gamma}\frac{\gamma!}{\beta!\alpha!}(\partial_z^{\beta}|z|^{-s})(\partial_z^{\alpha}|z|^{s})$$

we see that there remains only the term $\sigma(B)$ on the right and (15) follows.

1.6. In concluding this section we would like like to point out that after very minor adjustments all arguments here remain valid in a much more general situation as far as the spaces on which the operators act is corcerned. For example, we may replace the spaces L_s^2 by the L_s^p throughout provided that $1 < p < \infty$ (see [3]). For this purpose we merely have to use systematically Mihlin's multiplier theorem (see [7]) in estimating norms of operators in these spaces. Since the remaining results in this paper are mainly valid only in the L^2 case, we omit further details.

2. Local behaviour of singular integral operators. Asymptotic expansions in terms of differential operators with polynomial coefficients.

Singular integral operators have the property of acting like differential operators with polynomial coefficients on rapidly oscillating functions with small support. Specifically, let ν be unit vector and x_o a point in E^k and

$$H_t f = t^{-n/2} f\left(\frac{x-x_o}{t}+x_o\right)e^{-2\pi i(x\cdot\nu)/t^2}$$

then H_t is a unitary operator, and if f has compact support, the support of $H_t f$ shrinks to the point x_o as t tends to zero.

Let A be an operator in $\mathcal{S}_m$. Then

$$(16)\qquad H_t^* A H_t = \sum \frac{1}{\alpha!\,\beta!}\left[\partial_x^\alpha \partial_z^\beta \sigma(A)\right](x_o, \nu/t^2)\, t^{|\alpha|-|\beta|}\, x^\alpha \left(\frac{i}{2\pi}\frac{\partial}{\partial x}\right)^\beta + \rho_t$$

where the sum is extended over all terms of degree less than $2m$ in t and $\|\rho_t f\|_2 = 0(t^{2m})$ as t tends to zero for every f in the space $\mathcal{S}$ of rapidly decreasing infinitely differentiable functions.

The terms in the sum above are differential operators with polynomial coefficients multiplied by powers of t with non-negative exponents less than $2m$, the order of each operator not exceeding the corresponding exponent of t. This sum, therefore, is an operator depending on the parameter t and will be denoted by $\lambda(A)$. Evidently, $\lambda(A)$ is uniquely determined by A.

The mapping taking A into $\lambda(A)$ has the property that $\lambda(A^*) = \lambda(A)^*$ and $\lambda(A\,B) = \lambda(A)\,\lambda(B)$, if multiplication is effected modulo terms of degree larger than or equal to $2m$ in t. Therefore λ is a star representation of the algebra $\mathcal{S}_m$.

Let us prove this statement assuming that (16) holds. We have

$$H_t^* A\, B\, H_t\, f = (H_t^* A\, H_t)(H_t^* B\, H_t)\, f = (H_t^* A\, H_t)\left[\lambda(B) + \rho_t f\right].$$

Since $H_t^* A H_t$ is uniformly bounded in L^2 we have

$$\| H_t^* A\, H_t\, \rho_t\, f\|_2 \leq c\, \| \rho_t f\|_2 = 0(t^{2m})$$

and applying (16) to the coefficients of the various powers of t in

$\lambda(B) f$ we obtain

$$H_t^* A B H_t f = \lambda(A) \lambda(B) f + g_t = \lambda(A B) f + g_t'$$

where both $\|g_t\|_2$ and $\|g_t'\|_2$ are of the order t^{2m}, whence it follows that $\lambda(A B) = \lambda(A) \lambda(B)$.

On the other hand, from

$$H_t^* A H_t f = \lambda(A) f + h_t , \quad H_t^* A^* H_t g = \lambda(A^*) g + k_t$$

we obtain

$$(g, H_t^* A H_t f) = (g, \lambda(A) f) + (g, h_t)$$
$$(H_t^* A^* H_t g, f) = (\lambda(A^*) g, f) + (k_t, f)$$

and from the equality of the lefthand sides and the fact that the last terms on the right are of the order t^{2m} it follows that

$$(g, \lambda(A) f) = (\lambda(A^*) g, f)$$

and consequantly $\lambda(A^*) = \lambda(A)^*$.

Let us turn now to the proof of (16). We begin assuming that $x_0 = 0$ and that A is one of the operators $T_{n\ell}$ in (4). Let $\hat{H}_t$ be defined by $\hat{H}_t \hat{f} = (H_t f)^\wedge$

$$\hat{H}_t \hat{f} = t^{n/2} \hat{f}\left[t(z - \nu/t^2)\right], \quad \hat{H}_t^* f = t^{-n/2} f(z/t + \nu/t^2)$$

and setting $Y(z) = Y_{n\ell}(z)\, \varphi(z)\, |z|^{-d}$, $d = d_J$, we have

$$(H_t^* A H_t f)^\wedge = Y(z/t + \nu/t^2) \hat{f}(z) .$$

Replacing $Y(z)$ by its Taylor expansion at the point ν/t^2 we obtain

$$(H_t^* A H_t f)^\wedge = \sum \frac{1}{\alpha!} Y_\alpha(\nu/t^2)(z/t)^\alpha \hat{f}(z) + R(z/t, \nu/t^2) \hat{f}(z)$$

Let us split the sum here into two parts, one containing the terms with $|\alpha| < 2(m-d)$ and the other the remaining ones. Let g_t and h_t their inverse Fourier transforms respectively, and k_t the inverse Fourier transform of the last term on the right above. Then, as readily seen, $\lambda(T_{n\ell})f=g_t$, and we procede to estimate the norms of h_t and k_t. Using the estimates for the derivatives of $Y_{n\ell}(z)$ we see that for $t \leqslant 1$

$$\| h_t \|_2 \leqslant c\, t^{2m}\, n^{k+2m}\, \| f \|_{2m}$$

where the norm on the right is the norm of f as an element of L^2_{2m}. To estimate the norm of k_t we need estimates for the function $R(z/t, \nu/t^2)$. Let us set $r = 2m+1$ and estimate R for $t \leqslant 1/2$ and $|z| \leqslant t^{-1/r}$. We have

$$R(z/t, \nu/t^2) \leqslant c\, N\, |z|^r\, t^{-r}$$

where N is a bound for the derivatives of order r of $Y(z)$ in the sphere with center at ν/t^2 and radius $t^{-1-1/r}$, which is exterior to the sphere with center at the origin and radius $t^{-2}-t^{-1-1/r}$. Since $r \geqslant 3$ and $t \leqslant 1/2$ we see that $t^{-2}-t^{-1-1/r} \geqslant c\, t^{-2}$, $c > 0$, and on account of the estimates for the derivatives of $Y_{n\ell}(z)$ we have

$$N \leqslant c\, t^{2(d+r)}\, n^{k+2m}$$

and

$$R(z/t, \nu/t^2) \leqslant c\, |z|^r\, t^{2d+r}\, n^{k+2m} \leqslant c\, t^{2m}\, n^{k+2m}$$

On the other hand, for $|z| > t^{-1/r}$

$$R(z/t, \nu/t^2) \leqslant c\, |z|^r\, t^{-r}\, M \leqslant c\, |z|^r\, t^{-r}\, n^{k+2m}$$

where $M \leqslant c\, n^{k+2m}$ is a bound for the derivatives of order $r = 2m+1$ of $Y(z)$. From these two inequalities we find that

$$\|k_t\|_2 = \|R(z/t, \nu/t^2)\hat{f}(z)\|_2 \leq c\, t^{2m}\, n^{k+2m} \|\hat{f}\|_2 + c t^{-2m-1} n^{k+2m} \Big[\int_{|z|>t^{-1}/r} |z|^{2r} |\hat{f}(z)|^2 dz \Big]^{1},$$

But $\displaystyle\int_{|z|>t^{-1}/r} |z|^{2r} |\hat{f}(z)|^2 dz \leq t^{4r} \int_{|z|>t^{-1}/r} |z|^{2r+4r^2} |\hat{f}(z)|^2 dz \leq c\, t^{4r} \|f\|^2_{(4m+3)^2}$

where the norm on the right is the norm of f as an element of $L^2_{(4m+3)^2}$. Hence, substituting in the preceding estimate we get .

$$\|k_t\|_2 < c\, t^{2m}\, n^{k+2m} \|f\|_{(4m+3)^2}$$

Thus we have shown that

$$H_t^* T_{n\ell} H_t = \mathcal{A}(T_{n\ell}) + \rho_{n,t}$$

(17)

$$\|\rho_{n,t} f\|_2 \leq c\, t^{2m}\, n^{k+2m} \|f\|_{(4m+3)^2}$$

Let us assume now that A is the operator $A_{n\ell}$ in (4) . Let us set $a_{n\ell}(x) = a(x)$ and assume for simplicity that $x_0 = 0$. Then

$$H_t^* A_{n\ell} H_t f = a(tx)\, f(x) = \Big[\sum \frac{1}{\alpha!} a_\alpha(0)\, t^\alpha x^\alpha \Big] f(x) + R(tx)\, f(x)$$

where the sum is extended over all α with $|\alpha| < 2(m-d)$. The first term on the right is readily seen to be precisely $\mathcal{A}(A_{n\ell}) f$. To estimate the last term on the right we observe that for $t \leq 1$

$$|R(tx)| \leq c\, t^r |x|^r M \leq c\, t^{2(m-d)} |x|^r M$$

where r is the least integer larger than or equal to $2(m-d)$ and M is a bound for the derivatives of order r of $a(x)$. . From this we find that

$$H_t^* A_{n\ell} H_t = \mathcal{A}(A_{n\ell}) + \eta_{n\ell,t}$$

(18)

$$\|\eta_{n\ell,t} f\|_2 \leq c\, t^{2(m-d)} \|(1+|x|)^{2m} f(x)\|_2\, M_{n\ell}$$

where $M_{n\ell}$ is a bound for the derivatives of order less than or equal to $2m-[d]$ of $a_{n\ell}(x)$.

Let us turn now to the general case. Let A_j be the operator defined by one of the terms of the sum on the righthand side of (1). Then if the function $p_j(x,z)$ has the form $p_j(x,z)=q(x,z)\,\psi(z)\,|z|^{-d}$ we can expand A_j in series

$$A_j = \sum A_{n\ell}\, T_{n\ell} .$$

For each of the terms of the series we have

$$H_t^* A_{n\ell} T_{n\ell} H_t\, f = (H_t^* A_{n\ell} H_t)(H_t^* T_{n\ell} H_t) f = (H_t^* A_{n\ell} H_t)\left[\lambda(T_{n\ell}) f + \zeta_{n\ell,t} f\right] =$$

$$= \lambda(A_{n\ell})\lambda(T_{n\ell})\, f + \eta_{n\ell,t}\,\lambda(T_{n\ell})\, f + (H_t^* A_n H_t)\,\zeta_{n\ell,t}\, f$$

Now, on account of (3) we see that the norm of $A_{n\ell}$, which coincides with that of $H_t^* A_{n\ell} H_t$, is of the order n^{-r} for every r, and that the same estimate is valid for the numerical coefficients of $\lambda(A_{n\ell})$. On the other hand, the numerical coefficients of $\lambda(T_{n\ell})$ are of the order n^{k+2m}. Thus, using (17) and (18) and observing that $\lambda(A_{n\ell}).\lambda(T_{n\ell}) = \lambda(A_{n\ell} T_{n\ell})$ modulo terms of degree larger than or equal to $2m$ in t, summing the preceding equalities for all n and ℓ we obtain

$$H_t^* A_j H_t f = \lambda(A_j)\, f + g_t \ , \ \|g_t\|_2^2 = O(t^{2m})$$

Finally there remains to discuss the operator S in (1).

Since H_t^* is unitary and S maps L^2_{-m} continuously into L^2 we have

$$\| H_t^* S H_t f \|_2^2 = \|S H_t f\|_2^2 \leq c \int (1+|z|)^{-2m} \,|(H_t f)^\wedge(z)|^2\, dz$$

because the last integral is equivalent to the square of the norm of $H_t f$ as an element of L^2_{-m}. Now, $(H_t f)^\wedge = t^{n/2}\, \hat{f}\left[t(z-\nu/t^2)\right]$, whence substituting and setting $w = t(z - \nu/t^2)$ in the integral above we obtain

$$\|H_t^* S H_t f\|_2^2 \le c\int |\hat{f}(w)|^2 \left(1+|w/t+\nu/t^2|\right)^{-2m} dw$$

But, if $|w| \le 1/2t$ we have $|w/t+\zeta/t^2| \ge 1/2t^2$ and therefore

$$\|H_t^* S H_t f\|_2^2 \le c\int |\hat{f}(w)|^2 \left(1+1/2t^2\right)^{-2m} dw + c\int_{|w|>1/2t} |\hat{f}(w)|^2 dw =$$

$$= O(t^{4m}) + c\int_{|w|>1/2t} |\hat{f}(w)|^2 \, dw .$$

Now, $\hat{f}(w)$ is a rapidly decreasing function and therefore the last integral is $O(t^{4m})$. Thus we find that $\|H_t^* S H_t\|_2 = O(t^{2m})$, and the proof of (16) is complete, since the assumption $x_0=0$ is clearly irrelevant.

As an application of (16) let us show that $\sigma(A)$ is uniquely determined by A. Suppose that the lefthand side of (1) is zero and let $\sigma = \sum \tilde{p}_j(x,z)$ be the symbol associated with the righthand side. Then according to (16) we have

$$\lambda = \sum \frac{1}{\alpha!\,\beta!}\left(\partial_x^{\alpha}\partial_z^{\beta}\sigma\right)(x_0, \nu t^2)\, t^{|\alpha|-|\beta|}\, x^{\alpha}\left(\frac{i}{2\pi}\frac{\partial}{\partial x}\right)^{\beta} = 0 .$$

Let $\tilde{p}_l(x,z)$ be the first non-vanishing term in $\sigma = \sum \tilde{p}_j$ and let x_0 and ν be such that $\tilde{p}_l(x_0,\nu)\neq 0$. Then $\tilde{p}_l(x_0,\nu/t^2)$ is the term of lowest degree in t in the expression of λ. Since $\lambda = 0$ we must have $\tilde{p}_l(x_0,\nu/t^2)=0$, a contradiction. Consequently we must have $\sigma = 0$, as we wished to show.

3. Almost positive operators

3.1 Let us consider the operator Λ introduced in 1.5. Evidently, for $s \geq 0$, Λ^{-s} is positive in the sense that $(\Lambda^{-s} f, f) \geq 0$. Furthermore, if $s_1 \geq s_2$ then $\Lambda^{-s_1} \leq \Lambda^{-s_2}$. Let now A be a selfadjoint operator in $\mathcal{J}_m$. We will say that A is almost positive of order s, $0 \leq s \leq m$, on a set C if there exists a constant c such that

$$(A f, f) + c(\Lambda^{-s} f, f) \geq 0$$

for all f with support in C. Clearly, if A is almost positive of order s on a set C, then it is almost positive of order s' on C for every s', $s' \leq s$.

Given the symbol

$$\sigma(A) = \sum \tilde{p}_j(x, z)$$

of an operator in $\mathcal{J}_m$, the partial symbol $\sigma(A)]_s$ of order s, $0 \leq s \leq m$, of A will be the sum

$$\sigma(A)]_s = \sum \tilde{p}_j(x, z)$$

extended over all $\tilde{p}_j(x, z)$ of degree of homogeneity $-d_j$ in z with $-d_j > -s$.

Whether an operator A is almost positive of order s on C depends only on the behaviour of $\sigma(A)]_s$ in a neighborhood of C. More precisely,

<u>if A_1 is almost positive of order s on the set C and $\sigma(A_1)]_s = \sigma(A_2)]_s$ in an ε-neighborhood of C, then A_2 is also almost positive of order s on C.</u>

In fact, let $B = A_2 - A_1$. Then $\sigma(B)]_s = 0$ in an ε-neighborhood of C and if φ is an infinitely differentiable bounded function with bounded derivatives of all orders which equals 1 on C and vanishes with all its derivatives wherever $\sigma(B)]_s \neq 0$, we have

$$(A_2 f, f) = (A_1 f, f) + (B f, f) = (A_1 f, f) + (B \varphi f, \varphi f) =$$
$$= (A_1 f, f) + (\overline{\varphi} B \varphi f, f)$$

for every f with support in C. Now from formula (5) in 1.4 we see that $\sigma(\overline{\varphi} B \varphi)]_s = 0$, which according to what was said in 1.3 implies that $\Lambda^{s/2} \overline{\varphi} B \varphi \Lambda^{s/2}$ is bounded in L^2 Thus

$$|(\Lambda^{s/2} \overline{\varphi} B \varphi \Lambda^{s/2} g, g)| \leq c (g, g)$$

and setting $g = \Lambda^{-s/2} f$ and replacing above we obtain

$$(A_2 f, f) \geq (A_1 f, f) - c (\Lambda^{-s/2} f, \Lambda^{-s/2} f)$$
$$= (A_1 f, f) - c(\Lambda^{-s} f, f)$$

whence the desired conclusion follows.

Thus, almost positivity can be characterized in terms of partial symbols. Such a characterization, though, is not available at the presente time. Still, it is possible to give some non-trivial conditions on the partial symbols that imply almost positivity, and this will be the topic of this section.

3.2 We introduce now a special class $\mathcal{P}_m$ of symbols

$$\sigma = \sum \tilde{p}_j(x, z)$$

of operators in $\mathcal{S}_m$, satisfying the additional regularity condition that $\partial_x^\alpha \partial_z^\beta \tilde{p}_j(x, z)$ be continuous and bounded for $|\alpha| \leq 3m - 2d_j$, $|z| \geq 1$ and all β . Evidently, if A is in $\mathcal{S}_m$ then $\sigma(A)]_n$ belongs to $\mathcal{P}_n$ for $n \leq (2/3)m$.

We will define now an operation P_m on symbols in $\mathcal{P}_m$. It depends on the choice of a function φ in $\mathcal{S}$ and a non-negative infinitely differentiable function η on the real line with compact support in $t>0$. More precisely, the definition of P_m depends on the following constants

$$(19) \qquad c_{\alpha\beta} = \left(\frac{1}{2\pi i}\right)^{|\beta|} \int z^{\alpha} (\partial_z^{\beta} \varphi) \bar{\varphi} \, dz$$

$$(20) \qquad \eta_r = \int_0^{\infty} \eta(t) \, t^{r/2} \frac{dt}{t}$$

Given the symbol σ in $\mathcal{P}_m$, $P_m(\sigma)$ is defined by

$$P_m(\sigma) = \sum \frac{(-1)^{|\alpha+\beta|}}{\alpha!\,\beta!} c_{\alpha\beta} \eta_{|\alpha|-|\beta|} \partial_x^{\alpha} \partial_z^{\beta} \left[|z|^{\frac{|\beta|-|\alpha|}{2}} \sigma \right]$$

where summation extends over all terms of degree larger than $-m$ in z. Suppose we set $P_m = Q+R$, where $Q(\sigma) = c_{00}\eta_0\sigma$, and $R(\sigma)$ is the sum of the remaining terms in the preceding expression. Then R applied to a homogeneous terms of σ of degree $-d$ gives a sum of terms whose degree are at most $-d-(1/2)$, and consequently $R^{2m}=0$, and this clearly implies that P_m is invertible.

With these definitions we can state now our main result on almost positivity.

<u>Let A be a selfadjoint operator in $\mathcal{S}_n$, $m \leq (2/3)n$, and suppose that $P_m^{-1}[\sigma(A)](x,z) \geq 0$ for all x in an ε-neighborhood of the set C and $|z|$ sufficiently large. Then A is almost positive of order m in the set C.</u>

This is our basic criterion. Some generalizations of it will be given n 3.4.

The proof of our statement is rather lengthy. Let us set $P_m = P$ and begin showing that $P(\sigma)^{\#} = P(\bar{\sigma})$. This depends on certain relations between the $c_{\alpha\beta}$. Integrating by parts in (19) we obtain

$$c_{\alpha\beta} = \left(\frac{1}{2\pi i}\right)^{|\beta|}\int z^{\alpha}(\partial_z^{\beta}\varphi)\bar{\varphi}\,dz = \sum\left(\frac{i}{2\pi}\right)^{|\beta|}\frac{\alpha!\,\beta!}{\gamma!(\alpha-\gamma)!(\beta-\gamma)!}\int z^{\alpha-\gamma}(\partial_z^{\beta-\gamma}\bar{\varphi})\varphi\,dz =$$

$$= \pm\sum\left(\frac{i}{2\pi}\right)^{|\gamma|}\frac{\alpha!\,\beta!}{\gamma!(\alpha-\gamma)!(\beta-\gamma)!}\,\bar{c}_{\alpha-\gamma,\beta-\gamma}$$

Thus

$$P(\sigma)^{\#} = \sum\left(\frac{i}{2\pi}\right)^{|\gamma|}\frac{1}{\gamma!}\partial_x^{\gamma}\partial_z^{\gamma}\bar{P}(\sigma) =$$

$$= \sum\left(\frac{i}{2\pi}\right)^{|\gamma|}\frac{1}{\gamma!}\frac{(-1)^{|\alpha+\beta|}}{\alpha!\,\beta!}\bar{c}_{\alpha\beta}\,\eta_{|\alpha|-|\beta|}\,\partial_x^{\alpha+\gamma}\partial_z^{\beta+\gamma}\left[|z|^{\frac{|\beta|-|\alpha|}{2}}\bar{\sigma}\right]$$

$$= \sum\left(\frac{i}{2\pi}\right)^{|\gamma|}\frac{(-1)^{|\alpha+\beta|}}{\gamma!(\alpha-\gamma)!(\beta-\gamma)!}\bar{c}_{\alpha-\gamma,\beta-\gamma}\,\eta_{|\alpha|-|\beta|}\,\partial_x^{\alpha}\partial_z^{\beta}\left[|z|^{\frac{|\beta|-|\alpha|}{2}}\bar{\sigma}\right]$$

$$= \sum\frac{(-1)^{|\alpha+\beta|}}{\alpha!\,\beta!}c_{\alpha\beta}\,\eta_{|\alpha|-|\beta|}\,\partial_x^{\alpha}\partial_z^{\beta}\left[|z|^{\frac{|\beta|-|\alpha|}{2}}\bar{\sigma}\right]$$

as we asserted.

Next let us show that if σ is a homogeneous real non-negative symbol in $\mathcal{P}_m$, then there exists an operator A in $\mathcal{S}_m$ which is selfadjoint positive and such that $\sigma(A) = P(\sigma)$. Let $h(x,y)$ be the inverse Fourier transform of $\sigma(x,z)\eta(z)$, where η is the function in (20). Let $n > a > 0$ and consider the kernel

$$(21)\qquad k_n(x,y) = 2\int_a^n t^{3k-2d}\frac{dt}{t}\int\bar{\varphi}[t(x-w)]\,h[w,t^2(x-y)]\,\varphi[t(y-w)]\,dw$$

where $-d$ is the degree in z of σ and φ is the function in (19). Replacing above the Taylor expansions

$$\varphi[t(y-w)] = \sum t^{|\beta|}\frac{(y-x)^{\beta}}{\beta!}\varphi_{\beta}[t(x-w)] + R\,[t(x-w),\ t(y-w)]$$

$$h\,[w,t^2(x-y)] = \sum\frac{(w-x)^{\alpha}}{\alpha!}h_{\alpha}[x,t^2(x-y)] + T\,[x,w,t^2(x-y)]$$

where $|\alpha|$ and $|\beta|$ range from 0 to $[3m-2d]-1$, and setting

$$I_n(x,u) = \sum \frac{2}{\alpha!\,\beta!} \int_a^n t^{3k-2d} \frac{dt}{t} \int t^{|\beta|}(-u)^{\beta}(w-x)^{\alpha}\, \gamma[t(x-w)]\, \varphi_{\beta}[t(x-w)]\; h_{\alpha}(x, t^2u)\,dw$$

which, integrating with respect to w, becomes

$$I_n(x,u) = \frac{2(2\pi i)^{|\beta|}(-1)^{|\alpha+\beta|}}{\alpha!\,\beta!}\, c_{\alpha\beta} \int_a^n t^{2k-2d+|\beta|-|\alpha|}\, u^{\beta}\, h_{\alpha}(x, t^2u)\frac{dt}{t}$$

we obtain $k_n(x,y) = I_n(x, x-y) + S_n(x,y)$.

If we denote now by $p_n(x,z)$ the Fourier transform of $I_n(x,u)$ with respect to u we will have

$$\int p_n(x,z)\, e^{-2\pi i(x\cdot z)} \hat{f}(z)\,dz = \int f(y+x)\, I_n(x,-y)\,dy =$$

$$= \int I_n(x, x-y)\, f(y)\,dy$$

$$= \int k_n(x,y)\, f(y)\,dy - \int S_n(x,y) f(y)\,dy$$

Now, if

$$\psi_{\alpha\beta}(z) = 2\int_a^{\infty} \eta\left(\frac{|z|}{t^2}\right) t^{|\beta|-|\alpha|}\frac{dt}{t}$$

then $\psi_{\alpha\beta}(z)$ is a spherically symmetric function vanishing near the origin and coinciding with $\eta_{|\alpha|-|\beta|}|z|^{\frac{|\beta|-|\alpha|}{2}}$ for $|z|$ sufficiently large, and $p_n(x,z)$ converges boundedly to

$$p(x,z) = \sum \frac{(-1)^{|\alpha+\beta|}}{\alpha!\,\beta!}\, \partial_x^{\alpha}\, \partial_z^{\beta}\left[\sigma\, \psi_{\alpha\beta}(z)\right] \tag{22}$$

and consequently

$$\lim_{n\to\infty} \int \left[k_n(x,y) - S_n(x,y)\right] f(y)\,dy = \int p(x,z)\, e^{-2\pi i(x\cdot z)} \hat{f}(z)\,dz$$

$$= A_1 f$$

Now, if in the definition of $I_n(x,u)$ we restrict the summation to $|\alpha|+|\beta| < 2m-2d$, then A_1 will be an operator in $\mathcal{J}_m$ with $\sigma(A_1) = P(\sigma)$.

On the other hand the operator defined by the kernel $S_n(x,y)$ converges to an operator S with the property that S and $S(\frac{\partial}{\partial x})^{\alpha}, |\alpha| = m$ are bounded in L^2. Assuming this for the moment, this will imply that

$$A f = \lim_{n\to\infty} \int k_n(x,y)\, f(y)\, dy = A_1 f + S f$$

Now, since σ is real, $h(x,y) = \bar{h}(x,-y)$ and $k_n(x,y) = \bar{k}_n(y,x)$ and therefore $A_1 + S$ is selfadjoint. Consequently

$$S^* = A_1 - A_1^* + S$$

But $\sigma(A_1^*) = P(\sigma)^* = P(\bar{\sigma}) = P(\sigma) = \sigma(A_1)$ which implies that $A - A^*$ is in $\mathcal{J}_m$. Therefore we have that also

$$S^*(\frac{\partial}{\partial x})^{\alpha} = (A - A^*)(\frac{\partial}{\partial x})^{\alpha} + S(\frac{\partial}{\partial x})^{\alpha}, \quad |\alpha| = m$$

is bounded in L^2, and this imples that S is in $\mathcal{J}_m$, that A is also in $\mathcal{J}_m$ and $\sigma(A) = P(\sigma)$.

Let us return now to the operator S and show that it has the properties we claimed. The kernels $S_n(x,y)$ are linear combinations of the following functions

$$(23)\quad \int_a^n t^{3k-2d+|\beta|} (y-x)^{\beta}\, h_{\alpha}[x, t^2(x-y)] \frac{dt}{t} \int (w-x)^{\alpha} \varphi_{\beta}[t(x-w)]\, \bar{\varphi}[t(x-w)]\, dw$$

$$\int\int_a^n T[x, w, t^2(x-y)]\, \varphi[t(y-w)]\, \bar{\varphi}[t(x-w)]\, t^{3k-2d} \frac{dt}{t}\, dw$$

$$\int\int_a^n R[t(x-w), t(y-w)]\, (w-x)^{\alpha} \bar{\varphi}[t(x-w)]\, h_{\alpha}[x, t^2(x-y)]\, t^{3k-2d} \frac{dt}{t}$$

where in (23) $|\alpha + \beta| \geq 2m-2d$. Now, from the estimate

$$|\partial_y^{\alpha} R[t(x-w), t(y-w)]| \leq c\, t^{\ell} |x-y|^{\ell - |\alpha|}, \qquad 0 \leq |\alpha| \leq \ell$$

$$\left|\partial_y^{\alpha} T\left[x, w, t^2(x-y)\right]\right| \leq c \; t^{2|\alpha|} \, |w-x|^{\ell} \, (1+t^2 \, |x-y|)^{-3m-2k} \; , \quad 0 \leq |\alpha|$$

$$\left|\partial_y^{\beta} h_{\alpha}\left[x, t^2(x-y)\right]\right| \leq c \; t^{2|\beta|} \, (1+t^2 \, |x-y|)^{-3m-2k} \; , \quad 0 \leq |\beta|$$

where $\ell = [3m-2d]$, by substitution and integration above, we see that the kernels $S_n(x, y)$ converge and are dominated by an integrable function of $x-y$, which implies that the corresponding operators converge strongly to a limit S. Now, except for those kernels in (23) for which $|\alpha+\beta| = 2m-2d$, the same is true for their derivatives with respect to y of order m. This implies that the corresponding limit operators have the property of remaining bounded in L^2 after composition on the right with differentiation of order m. Thus we only have to verify still that the limits of the operators defined by the kernels in (23) also remain bounded after composition on the right with differentiation of order m. For this purpose we express these operators by means of the Fourier transforms of the functions on which they operate and obtain the expression

$$c \int \partial_x^{\alpha} \partial_z^{\beta}\left[\sigma \psi_{\alpha\beta}(z)\right] (2\pi iz)^{\gamma} \, e^{-2\pi i(x-z)} \hat{f}(z) \, dz$$

for the limit operators acting on $\partial_x^{\gamma} f(x)$. Now, if $m=|\gamma|$ and $m \geq 2$ the preceding integral is readily seen to represent an operator in $\mathcal{O}_n$, $2n<m$, acting on f . Since according to 1.3 this operator is bounded in L^2 , the desired conclusion follows. If m=1, the function $\partial_x^{\alpha} \partial_z^{\beta}\left[\sigma \psi_{\alpha\beta}(z)\right](2\pi iz)^{\gamma}$, though still infinitely differentiable in z, is not sufficiently regular in x to belong to any $\mathcal{O}_n$. However, the reader will easily verify that the argument employed in 1.3 still applies in this case, and the desired conclusion again follows.

Summarizing, we have shown that the kernels (21) define operators converging to a selfadjoint operator A in $\mathcal{O}_m$ with $\sigma(A) = P(\sigma)$.

Let us show now that A is positive. Let the operators $A(w, t)$ and $\Phi(w,t)$ be defined by

$$(24)\quad A(w,t)f = \int h\left[w, t^2(x-y)\right] t^{2k-2d} f(y)\, dy = \int \sigma(w,z)\,\eta\left(\frac{z}{t^2}\right) e^{-2\pi i(x.z)}\hat{f}(z)dz$$

$$\Phi(w,t)f = \varphi\left[t(x-w)\right] t^{k/2} f(x)$$

Then, as readily seen,

$$\int k_n(x,y)\, f(y)\, dy = \int\int_a^n \Phi^*(w,t)\, A(w,t)\, \Phi(w,t)\, f \;\frac{dt}{t}\, dw$$

Now, since σ and η are real non-negative, $A(w,t)$ is selfadjoint positive, and the same is true of the operator defined by the kernel $k_n(x,y)$. Passing to the limit, we find that A is positive.

Now we generalize this result. Let $\sigma(x,z)$ be the symbol of an operator in $\mathcal{P}_m$ such that $\sigma(x,z) \geq 0$ for $|z| \geq N$. Let $A(w,t)$ be defined in terms of σ as in (24). Then the operator

$$\int\int_a^n \Phi^*(w,t)\, A(w,t)\, \Phi(w,t)\;\frac{dt}{t}\, dw$$

is positive provided that $\eta(|z|/t^2) = 0$ for $|z| \leq N$ and $t \geq a$, which is true for a sufficiently large. Applying our preceding results to each homogeneous term of σ we see that the integral above converges to an operator A in $\mathcal{S}_m$ with $\sigma(A) = P(\sigma)$ which, evidently, is positive.

Finally, let A be an operator in $\mathcal{S}_n$ which is selfadjoint and such that for some integer m, $m \leq (2/3)n$, $|z| \geq N$, we have $P^{-1}\{\sigma(A)\}$ $(x,z) \geq 0$ for all x in an ε-neighborhood of a set C. Let φ be a non-negative bounded function with bounded derivatives of all orders which equals 1 outside that neighborhood and vanishes in another, smaller ε-neighborhood of C. Since $P^{-1}\{\sigma(A)\}$ is a bounded function, for c positive and sufficiently large, $P^{-1}\{\sigma(A)\} + c\,\varphi$ will be non-negative for $|z| \geq N$ and all x. Consequently, there will exist a positive operator B in $\mathcal{S}_m$ with

$$\sigma(B) = P\left[P^{-1}\{\sigma(A)\} + c\varphi\right] = \sigma(A) + cP\varphi$$

Since $P\varphi$ vanishes in a neighborhood of C, we find that $\sigma(B) \doteq \sigma(A)$ in that same $\mathcal{E}$-neighborhood. Consequently, since B is positive, A is almost positive of order m. This concludes the proof of our main result.

3.3 In this paragraph, with applications in mind, we will make a specific choice of the function φ in (19) and we will calculate the operators P and P^{-1} more explicitly. For φ we will take the function $2^{k/4}\exp(-\pi|z|^2)$ and we will calculate the coefficients $c_{\alpha,\beta}$ by means of the generating function

$$F(x,y) = \sum \frac{(-1)^{|\alpha+\beta|}}{\alpha!\,\beta!} c_{\alpha,\beta}\, x^{\alpha} y^{\beta}.$$

In order to find $F(x,y)$ we substitute the $c_{\alpha,\beta}$ by their expression in (19) and obtain

$$F(x,y) = 2^{k/2}\int\Big(\sum \frac{(-1)^{|\alpha|}}{\alpha!} x^{\alpha} z^{\alpha} e^{-\pi|z|^2}\Big)\Big(\sum \frac{(-1)^{|\beta|}}{\beta!}\Big(\frac{1}{2\pi i}\Big)^{|\beta|} y^{\beta}\, \partial_z^{\beta}\, e^{-\pi|z|^2}\Big)dz$$

Now, the first parenthesis under the integral sign equals $\exp\left[-(x.z)-\pi|z|^2\right]$ and the second is the Taylor expansion of $\exp(-\pi|z|^2)$ at the point z and therefore equals $\exp\left[-\pi|z-y/2\pi i|^2\right]$. Thus, we have

$$F(x,y) = 2^{n/2}\int \exp\left[-\pi|z|^2-(x\cdot z)-\pi|z-y/2\pi i|^2\right]dz$$

$$= \exp\left[\frac{1}{8\pi}(|x|^2+2i(x\cdot y)+|y|^2)\right] =$$

$$= \sum \frac{1}{n!}\frac{1}{m!}\frac{1}{\ell!}\Big(\frac{1}{8\pi}\Big)^{n+m+\ell}(x.x)^n(y.y)^m(2i(x\cdot y))^{\ell}$$

In order to further simplify the expression of P we choose now η in (20) in such a way that $\eta(e^s)$ be an even function of s. With such a choice we will have $\eta_r = \eta_{-r}$, and we normalize η so that $\eta_0 = 1$. Letting now Δ_x and Δ_z denote the Laplacians with respect to x and z, and setting

$$\nabla_x \cdot \nabla_z = \sum \frac{\partial}{\partial x_j} \frac{\partial}{\partial z_j}\ .$$

we have the following formal expression for P

$$P = \sum \frac{1}{n!} \frac{1}{m!} \frac{1}{\ell!} \left(\frac{1}{8\pi}\right)^{n+m+\ell} \eta_{2(n-m)} \Delta_x^n \Delta_z^m (2i \nabla_x \cdot \nabla_z)^\ell |z|^{m-n}$$

Now let us calculate P and P^{-1} in some of the classes $\mathcal{P}_m$.

If $m=1$, then $P_1 = P_1^{-1} = 1$.

If $m=2$

$$P_2 = 1 + \frac{1}{8\pi}\left[\eta_2(\Delta_x |z|^{-1} + \Delta_z |z|) + 2i(\nabla_x \cdot \nabla_z)\right]$$

$$P_2^{-1} = 1 - \frac{1}{8\pi}\left[\eta_2(\Delta_x |z|^{-1} + \Delta_z |z|) + 2i(\nabla_x \cdot \nabla_z)\right]$$

If $m=3$

$$P_3 = 1 + \frac{1}{8\pi}\left[\eta_2(\Delta_x |z|^{-1} + \Delta_z |z|) + 2i(\nabla_x \cdot \nabla_z)\right] +$$

$$+ \left(\frac{1}{8\pi}\right)^2 \left[\frac{\eta_4}{2}(\Delta_x^2 |z|^{-2} + \Delta_z^2 |z|^2) - 2(\nabla_x \cdot \nabla_z)^2 + \Delta_x \Delta_z +\right.$$

$$\left. + 2i\eta_2 (\nabla_x \cdot \nabla_z)(\Delta_x |z|^{-1} + \Delta_z |z|)\right]$$

$$P_3^{-1} = 2 - P_3 + \left(\frac{1}{8\pi}\right)^2 \left[\eta_2(\Delta_x |z|^{-1} + \Delta_z |z|) + 2i(\nabla_x \cdot \nabla_z)\right]^2$$

3.4 The results on almost positivity we have obtained so far can be extended to matrices of singular integral operators and, more generally, pseudo-differential operators.

Let us consider square matrices of operators in $\mathcal{S}_m$ and regard them as operators acting on vector-valued functions. Such functions will be said to belong to L^2_s, $s \geq 0$, if their components belong to L^2_s. A vector-valued distribution in L^2_s, $s < 0$, is simply a vector whose components are distributions in L^2_s. The inner product of two vector-valued functions in L^2 is the integral of their pointwise inner products. With these definitions all results obtained in section 1 are seen to generalize, replacing symbols by the corresponding matrices and their complex conjugates by the corresponding adjoints. All definitions and results in sections 2 and 3 generalize as well, and in (19) we can replace φ by a matrix-valued function replacing at the time $\overline{\varphi}$ by the adjoint of the latter, provided the resulting matrix c_{oo} is non-singular.

We can also extend the notion of almostpositivity to pseudo-differential and matrices of pseudo-differential operators as follows. Let P be a self-adjoint pseudo-differential operator of order t; we will say that P is almost positive of order s on a set C if there exists a constant c such that $(P f, f)+c(\Lambda^{t-s} f, f) \geq 0$ for all infinitely differentiable functions f with support in C. If P is a mixed pseudo-differential operator and f is a function in $\mathcal{S}$ then P f is a tempered distribution and (P f, f) is meaningful as the distribution P f evaluated on the testing function $\overline{f}$. Of course, an analogous definition is meaningful for matrices of pseudo-differential operators. Now we have the following result which can be combined with the results of 3.3 to obtain criteria for almost positivity.

Let A and B singular integral operators in $\mathcal{S}_m$. Let A be self-

adjoint and suppose that $\Lambda^t A \Lambda^t$ $-\infty<t<\infty$, is almost positive of order s , $s\leq m$, in an $\mathcal{E}$-neighbourhood of the set C . Then $\Lambda^r B^* A B \Lambda^r$ is almost positive of order s in C.
Conversely, if $\Lambda^r B^* A B \Lambda^r$ is almost positive of order s in an $\mathcal{E}$-neighbourhood of the set C , and if the principal symbol of B , that is the homogeneous term of degree zero of $\sigma(B)$, is bounded away from zero in this neighbourhood , then $\Lambda^t A \Lambda^t$ is álmost positive of order s in C .

The same result holds for matrices of operators with the only difference that in the second part it is the determinant of the principal symbol of B which must be bounded away from zero in an $\mathcal{E}$-neighbourhood of C .

The validity of our statement follows from a number of observations which we list below.

a) Suppose that A is in $\mathcal{J}_m$ and that $|t|\leq 2m$. Then $A^*\Lambda^t A \leq c\Lambda^t$. In fact, let $B=\Lambda^{-t/2} A^*\Lambda^t A\Lambda^{-t/2}$. Then since Λ^r maps L^2_t continuously into L^2_{t-r}, it follows from what was seen in 1.3 that B is bounded in L^2 . Thus $A^*\Lambda^t A = \Lambda^{t/2} B \Lambda^{t/2} \leq c\Lambda^t$.

b) Suppose that $\Lambda^t A \Lambda^t$ is almost positive of order s in C and that φ is bounded, has infinitely many bounded derivatives, and is supported by C . Then

$$\bar{\varphi}\Lambda^t A\Lambda^t\varphi \geq -c\Lambda^{2t-s}$$

Indeed, we have

$$(\bar{\varphi}\Lambda^t A \Lambda^t\varphi f, f) = (\Lambda^t A \Lambda^t\varphi f, \varphi f) \geq -c(\Lambda^{2t-s}\varphi f, \varphi f) =$$

$$= -c(\bar{\varphi}\Lambda^{2t-s}\varphi f, f)$$

Thus , according to a)

$$\bar{\varphi}\Lambda^t A\Lambda^t\varphi \geq -c\bar{\varphi}\Lambda^{2t-s} \geq -c\Lambda^{2t-s}$$

c) Suppose that $\bar{\varphi}\Lambda^t A\Lambda^t\varphi \geq -c\Lambda^{2t-s}$, where φ is bounded and has infinitely many bounded derivatives and equals 1 on the set C , then $\Lambda^t A\Lambda^t$ is almost positive of order s in C.

This is evident.

d) Suppose that $\Lambda^t A_1\Lambda^t$ is almost positive of order s on an ε-neighbourhood C_ε of C and that $\sigma(A_2) = \sigma(A_1)$ in C_ε. Then $\Lambda^t A_2\Lambda^t$ is almost of order s in C, provided that A_1 and A_2 belong to $\mathcal{S}_m$, $m \geq s$. Let $B = A_2 - A_1$ and φ a bounded funct on with infinitely many bounded derivatives which equals 1 on C and vanishes outside C_ε. Then $\sigma(B) = 0$ in C_ε and

$$\bar{\varphi}\Lambda^t A_2\Lambda^t\varphi = \bar{\varphi}\Lambda^t A_1\Lambda^t\varphi + \bar{\varphi}\Lambda^t B\Lambda^t\varphi \geq -c\Lambda^{2t-s} + \bar{\varphi}\Lambda^t B\Lambda^t\varphi$$

But since $\sigma(B)$ vanishes in C_ε and φ vanishes outside C_ε, according to what was seen in 1.5 $\bar{\varphi}\Lambda^t B\Lambda^t\varphi = \Lambda^t B_1\Lambda^t$ with B_1 in $\mathcal{S}_m$ and $\sigma(B_1) = 0$. Consequently $\Lambda^{s/2} B_1\Lambda^{s/2}$ is bounded in L^2 and

$$\Lambda^t B_1\Lambda^t = \Lambda^{t-s/2}(\Lambda^{s/2} B_1\Lambda^{s/2})\Lambda^{t-s/2} \leq c\Lambda^{2t-s}$$

whence it follows that $\bar{\varphi}\Lambda^t A_2\Lambda^t\varphi \geq -c\Lambda^{2t-s}$. Since $\varphi = 1$ on C, according to c) $\Lambda^t A_2\Lambda^t$ is almost positive of order s on C .

e) Suppose that A is in $\mathcal{S}_m$ and that $\Lambda^t A\Lambda^t$ is almost positive of order s, $s \leq m$, in an ε-neighbourhood of the set C . Then $\Lambda^r A\Lambda^r$ is almost positive of order s in C .

Let C_ε be the given ε-neighbourhood of C a d let φ be bounded and have infinitely many bounded derivatives and be equals to 1 on and vanish outside $\mathcal{O}_\varepsilon$. Then, according to b),

$$\bar{\varphi}\Lambda^t A \Lambda^t \varphi \geq -c\Lambda^{2t-s}$$

whence it follows that

$$\Lambda^r(\Lambda^{-t}\bar{\varphi}\Lambda^t A\Lambda^t\varphi\Lambda^{-t})\Lambda^r = \Lambda^r B\Lambda^r \geq -c\Lambda^{2r-s}$$

But, since $\varphi=1$ in $\mathcal{O}_{\varepsilon/2}$, $\sigma(B)=\sigma(A)$ in $\mathcal{O}_{\varepsilon/2}$, and since $\Lambda^r B\Lambda^r$ is almost positive of order s, according to d) $\Lambda^r A\Lambda^r$ is almost positive of order s in C.

f) Suppose that A and B are in $\mathcal{S}_m$, and that A is selfadjoint and almost positive of order s, $s\leq m$, in an ε-neighborhood of the set C, then $B^* A B$ is almost positive of order s in C.

Let φ be bounded and have infinitely many bounded derivatives and be equal to 1 in $\mathcal{O}_{\varepsilon/2}$ and vanish outside $\mathcal{O}_\varepsilon$. Then, according to b) $\bar{\varphi}A\varphi \geq -c\Lambda^{-s}$. On the other hand, according to a)

$$B^*\bar{\varphi}A\varphi B \geq -c\, B^*\Lambda^{-s}B \geq -c\Lambda^{-s}$$

and thus $B^*\bar{\varphi}A\varphi B$ is almost positive of order s. But $\sigma(B^*\bar{\varphi}A\varphi B) = \sigma(B^* AB)$ in $\mathcal{O}_{\varepsilon/2}$, and consequently, according to d), B^*AB is almost positive of order s in C.

g) Suppose that A and B are in $\mathcal{S}_m$, that A is selfadjoint and that $B^* A B$ is almost positive of order s, $s\leq m$, in an ε-neighbourhood $\mathcal{O}_\varepsilon$ of the set C. Suppose furthermore that the principal symbol of B is bounded away from zero in $\mathcal{O}_\varepsilon$. Then A is almost positive of order s in C.

Let B_1 be in $\mathcal{S}_m$ and let its principal symbol coincide with the inverse of the principal symbol of B in $\mathcal{O}_\varepsilon$. Then using (5) we see that $B B_1 = I - R$, where I is the identity operator and the principal symbol of R vanishes in $\mathcal{C}_\varepsilon$. Now, since the principal symbol of R vanishes in $\mathcal{C}_\varepsilon$, using (5) aga n, we see that for n sufficiently large $\sigma(R^n)=0$ in $\mathcal{C}_\varepsilon$. Consequently, if $B_2=B_1(I+R+R+\ldots+R^{n-1})$, we find that $\sigma(BB_2)=\sigma(I-R^n)=1$ in $\mathcal{C}_\varepsilon$. But B^*AB is almost positive of order s in $\mathcal{C}_\varepsilon$, and, according to f), this implies that $B_2^*B^*ABB_2$ is almost positive of order s in $\mathcal{C}_{\varepsilon/2}$. On the other hand, since $\sigma(BB_2)=1$ in $\mathcal{C}_\varepsilon$, we have $\sigma(B_2^*B^*ABB_2)=\sigma(A)$ in $\mathcal{C}_\varepsilon$, and, according to d), it follows that A is almost positive of order s in C.

Finally, let us suppose that $\Lambda^t A \Lambda^t$, A in $\mathcal{S}_m$, is almost positive of order s in an ε-neighborhood $\mathcal{C}_\varepsilon$ of C. Then from e), f) and e) again, it follows that A, B^*AB and $\Lambda^r B^*AB\Lambda^r$ are almost positive of order s in $\mathcal{C}_{2/3\varepsilon}$, $\mathcal{C}_{\varepsilon/3}$, and C respect vely.
If we assume on the other hand that $\Lambda^r B^*AB\Lambda^r$ is almost positive in $\mathcal{C}_\varepsilon$ and that the principal symbol of B does not vanish there, the almost positivity of $\Lambda^t A \Lambda^t$ in C follows by applying e) g) and e) successively.

To conclude this section we will prove one more criterion for almost positivity which is useful in applications.

Let A and A_1 be selfadjoint operators in $\mathcal{S}_m$. Let $0<s\leq[\frac{2}{3}m]$. Suppose that A is almost positive of order s in an ε-neighbourhood $\mathcal{C}$ of the set C and that the principal symbol p of A_1 is non-negative in $\mathcal{C}$. Let $\mathcal{C}_1$ be the set of points (x, z) such that x is in $\mathcal{C}$ and $p<\varepsilon$, $\varepsilon>0$. Then if

$$P_n^{-1}[\sigma(A_1-A)] \geq -c\,|z|^{-s}, \quad n=\left[\frac{2}{3}m\right]$$

for all (x, z) in $\mathcal{C}_1$ with $|z|$ sufficiently large, A_1 is almost positive of order s in C.

Let $\varphi_j(t)$ be defined in $0 \leq t < \infty$, be infinitely differentiable and have the properties $0 \leq \varphi_j \leq 1$, $\varphi_j = 0$ for $t \leq \varepsilon/2^j$, $\varphi_j = 1$ for $t \geq \varepsilon/2^{j-1}$, and let $\mathcal{C}_o = \{(x, z) \mid x \in \mathcal{O}\}$, $\mathcal{C}_j = \{(x, z) \mid x \in \mathcal{C}, p < \varepsilon/2^{j-1}$. Then $\varphi_j(p) = 1$ in $\mathcal{C}_o - \mathcal{C}_j$ and $\varphi_j(p) = 0$ in $\mathcal{C}_{j+1}$. Define K_j by $\sigma(K_j) = \varphi_j(p)$.

Let $0 \leq s_1 < s_2 < \dots$ be the elements of the semigroup generated by the numbers delete, n and minus the degrees of homogeneity in z of the terms in $\sigma(A)$ and $\sigma(A_1)$. Let

$$A_{j+1} = (1 + c_j K_j \Lambda^{-s_j})^* A_j (1 + c_j K_j \Lambda^{-s_j}), \qquad h_j = \sigma(A_j - A), \; j = 1, 2, \dots, \; P = P_n$$

and we will show that with an appropriate choice of the constants $c_j > 0$ and $c_j' > 0$ the h_j will satisfy the inequalities

$$
\begin{aligned}
& P^{-1}(h_j) \geq -c\,|z|^{-s} && \text{in } \mathcal{C}_o - \mathcal{C}_{j-1} \\
\text{(25)} \qquad & P^{-1}(h_j) \geq P^{-1}(h_1) - c_j'\,|z|^{-s_j} && \text{in } \mathcal{C}_{j-1} - \mathcal{C}_j \\
& P^{-1}(h_j) = P^{-1}(h_1) && \text{in } \mathcal{C}_j\,.
\end{aligned}
$$

for $|z|$ sufficiently large. For $j = 1$ this is clearly true. Assuming, as we may, that in (19) and (20) we have $c_{oo} \cdot \eta_o = 1$, and denoting by p_j the principal symbol of A_j, from the inductive definition of A_j and h_j one sees readily that

$$\text{(26)} \qquad P^{-1}(h_{j+1}) = P^{-1}(h_j) + 2c_j p_j \varphi_j(p)\,|z|^{-s_{j}} + q_j(x, z)$$

where $q_j(x, z) = \varphi_j(p) = 0$ in $\mathcal{O}_{j+1}$ and $q_j(x, z)$ is a sum of homogeneous terms of degrees less than or equal to $-s_{j+1}$ in z and therefore satisfies the inequality $|q_j(x, z)| \leq c_{j+1}'\,|z|^{-s_{j+1}}$ for $|z| \geq 1$. Thus, assuming that (25) holds for h_j we will have

$$P^{-1}(h_{j+1}) = P^{-1}(h_j) = P^{-1}(h_1) \text{ in } \mathcal{O}_{j+1}$$

and, since as readly seen $p_j \gtrsim p \gtrsim 0$ in $\mathcal{O}_o$

$$P^{-1}(h_{j+1}) \gtrsim P^{-1}(h_j) - c'_{j+1} |z|^{-s_{j+1}} = P^{-1}(h) - c'_{j+1} |z|^{-s_{j+1}} \text{ in } \mathcal{O}_j - \mathcal{O}_{j+1}$$

and the two last condition in (25) are also satisfied by h_{j+1}. To see that h_{j+1} satisfies the first as well observe that in $\mathcal{O}_o - \mathcal{O}_j$ we have $\varphi_j(p) = 1$ and $p_j \gtrsim p\ \varepsilon/2^{j-1}$. Thus, taking c_j so that $2c_j(\varepsilon'/2^{j-1}) \gtrsim c'_j + 1$, (26) yields

$$P^{-1}(h_{j+1}) \gtrsim P^{-1}(h_j) + (c'_j + 1)|z|^{-s_j} - c'_{j+1}|z|^{-s_{j+1}}$$

which implies that

$$P^{-1}(h_{j+1}) \gtrsim P^{-1}(h_j) + c'_j |z|^{-s_j} \text{ in } \mathcal{O}_o - \mathcal{O}_j$$

for $|z|$ sufficiently large. If $j = 1$, this inequality reduces to $P^{-1}(h_2) \gtrsim P^{-1}(h_1) + c'_1$, in $\mathcal{O}_o - \mathcal{O}_1$. Taking c'_1 sufficiently large we see that the first inequality in (25) is satisfied for $j = 2$. If on the other hand $j > 1$ then according to our hypotheses we have $P^{-1}(h_1) \gtrsim -c|z|^{-s}$ in $\mathcal{O}_1$ and, consequently, from the first two inequalities in (25) it follows that $P^{-1}(h_j) \gtrsim -c|z|^{-s} - c'_j |z|^{-s_j}$ in $\mathcal{O}_o - \mathcal{O}_j$. Thus substituting above we obtain

$$P^{1}(h_{j+1}) \gtrsim -c|z|^{-s} \text{ in } \mathcal{O}_o - \mathcal{O}_j$$

and h_{j+1} also satisfies (25), which we see now to be valid for all j. Now, if $s_j = n$, since $s_1 = 0$, we must have $j \gtrsim 2$ and therefore $\mathcal{O}_{j-1} \subset \mathcal{O}_1$. Thus $P^{-1}(h_1) \gtrsim -c|z|^{-s}$ in $\mathcal{O}_{j-1}$, and (25) implies that

$$P^{-1}(h_j) \gtrsim -c|z|^{-s} + c'_j |z|^{-n}, \quad c > 0$$

in $\mathcal{O}_o$. But since $s < n$, this in turn implies that

$$P^{-1}(h_j) \gtrsim -c|z|^{-s}, \quad |z| \geq 1 .$$

in $\mathcal{O}_o$ for c sufficiently large. Now, according to the results of 3.2 this implies that $A_j - A + c\Lambda^{-s}$ is almost positive of order n in an $(\varepsilon/2)$-neighbourhood $\mathcal{O}'$ of C, for c sufficiently large and since A is almost positive of order s in $\mathcal{C}$, the same holds for $A_j + c\Lambda^{-s}$ and A_j. But if $B = \prod_1^j (1+c_i \dot{K}_i \Lambda^{-s_i})$, then the principal symbol of B is larger than or equal to 1 and $A_j = B^* A_1 B$, and the preceding result allows us to to conclude that A_1 is almost positive of order s in C.

4. A priori inequalities for singular integral operators

In this section we shall be concerned with the validity of a priori inequalities for singular integral operators and the related problem of existence of solutions of equations of the form

$$A f = g \qquad (27)$$

where A is a pseudo-differential operator and g is an element of L^2_r. We shall limit ourselves to describing some general methods to discuss the inequalities and related existence theorems and illustrate them with some simple examples.

4.1 Let $\mathcal{O}$ be a bounded open subset of E^k, R the operation of restricting functions and distributions to $\mathcal{O}$, and $L^2_u(\mathcal{O})$ the image of L^2_u under R, with the quotient space norm.
Let $A = \Lambda^t B \Lambda^r$ with B in $\mathcal{S}_m$, $|u| < m-1$, $0 \leq 2s < [m - |u|]$. Then if for some c, $c>0$, $T T^* - c\Lambda^{-2s}$, $T = \Lambda^u B \Lambda^{-u}$, is almost positive of order s', $2s < s'$, in an ε-neighbourhood of $\mathcal{O}$, the image of L^2_{u+r} under R A intersected with $L^2_{u-t+s}(\mathcal{O})$ is a closed subspace of

finite codimension of $L_{u-t+s}(\mathcal{O})$.
Conversely, if the last proposition holds and C is any set at a positive distance from the complement of $\mathcal{O}$, there exist s', $2s<s'$, and c, $c>0$, such that $T T^* - c\Lambda^{-2s}$ is almost positive of order s' in C.

Let A' be the restriction of RA to the subspace of L^2_{u+r} consisting of the elements that RA maps into $L^2_{u-t+s}(\mathcal{O})$. Then A' as an operator from L^2_{u+r} into $L^2_{u-t+s}(\mathcal{O})$ is evidently closed and its range is precisely the intersection of the image of L^2_{u+r} under RA with $L^2_{u-t+s}(\mathcal{O})$. Now, the dual of $L^2_{u-t+s}(\mathcal{O})$ is the closure of $C^\infty_0(\mathcal{O})$ in L^2_{-u+t-s}, and, as is well known, the range of A is closed and of finite codimension if and only if its adjoint, which coincides with A^* in $C^\infty_0(\mathcal{O})$, satisfies the inequality

$$c(\|f\|^2_{-u+t-s'/2}+\|A^*f\|^2_{-u-r}) > \|f\|^2_{-u+t-s}$$

for all f in $C^\infty_0(\mathcal{O})$, with $\|f\|_u$ denoting the norm of f as an element of L^2_u. But $\|f\|^2_u$ is equivalent with $(\Lambda^{2u}f, f)$, and since $B\Lambda^{-u} = \Lambda^{-u}T^u$, substituting, we see that the inequality above is equivalent with

$$c_1\Lambda^{-2u+2t-s'}+\Lambda^{t-u}\, T\, T^*\Lambda^{t-u} - \frac{1}{c}\Lambda^{-2u+2t-2s} > 0, \quad c>0$$

whence it follows that the range of A' is closed and of finite codimension if and only if

$$\Lambda^{t-u}(T\, T^* - \frac{1}{c}\Lambda^{-2s})\Lambda^{t-u}$$

is almost positive of order s' in $\mathcal{O}$, and our statement follows now from the results of 3.4.

In order to establish the almost positivity of $T\,T^* - c\Lambda^{-2s}$ on a given set we could use the results of section 3. However, due to the special form of the operator under consideration, such a direct approach does not lead to satisfactory results. For example, suppose that $\sigma(T)$ is a numerical (as opposed to matrix valued) function and assume that $\sigma(T)(x_0, z_0)=0$ while not all first order derivatives of $\sigma(T)$ vanish at (x_0, z_0). Then one can easily see that choosing φ in (19) as in 3.3

$$P^{-1}\left[\sigma(T\ T^*)\right](x_0, tz_0) \sim -c\,t^{-1}, \quad c>0$$

as $t \to +\infty$. This shows that to obtain any positive results by such a direct approach one must adapt the choice of φ to the nature of T. Our chances of success improve though if we use the following result.

Let T, A, A_1 and B be operators in $\mathcal{J}_m$ and suppose that the principal symbol of A is bounded away from zero in an ε-neighbourhood $\mathcal{C}$ of the set C. Let $T_1 = A\,T\,A_1$ and $H = T_1 T_1^* - B^*B$ and denote by p the principal symbol of $T_1 T_1^*$ and by $\mathcal{S}_1$ the set of points (x, z) such that x is in $\mathcal{C}$ and $p(x, z) < \varepsilon, \varepsilon > 0$. Then if

$$P_n^{-1}\left[\sigma(H)\right] \geq c\,|z|^{-2s}, \qquad c>0$$

for $0 \leq 2s < n = \left[\frac{2}{3} m\right]$ and all (x, z) in $\mathcal{S}_1$ with $|z|$ sufficiently large, $T\,T^* - c\Lambda^{-2s}$ is almost positive of order n in C for some positive c.

The advantage here is that, in addition to the choice of φ in (19), we have now the choice of A, A_1 and B. On the other hand, the proof of this is very simple. In fact, according to the last theorem in 3.4 our hypotheses imply that $T_1 T_1^* - c'\Lambda^{-2s}$, $c'>0$, is almost positive of order n in an $(\varepsilon/2)$-neighbourhood $\mathcal{C}'$ of the set C. On the other hand

$$T_1 T_1^* \leq c\,A\,T\,T^* A^*$$

and, according to a) in 3.4

$$A \Lambda^{-2s} A^* \leq c \Lambda^{-2s}$$

Thus , for c_1 sufficiently large we will have

$$T_1 T_1^* - c' \Lambda^{-2s} \leq c_1 A T T^* A^* - A \Lambda^{-2s} A^*$$

and consequently $A (C T T^* - \Lambda^{-2s}) A^*$ is almost positive of order n in $\mathcal{C}_1$ and the desired result follows.

4.2 In this paragraph we will make some more concrete applications of the results obtained so far. We will assume that the operator T associated with equation (27) has the property that $\sigma(T)$ consists only of terms which are homogeneous of negative integral degree in z. This is the case, for example , if (27) is a differential equation. Also, we shall use the following abbreviation

$$[p, q] = \frac{i}{2\pi} \sum \frac{\partial p}{\partial z_j} \frac{\partial q}{\partial x_j} - \frac{\partial q}{\partial z_j} \frac{\partial p}{\partial x_j}$$

Let T be an operator in $\mathcal{S}_m$, $p = \sigma(T)$ and p_o its principal symbol. Let C be a set, $\mathcal{C}$ and $\mathcal{E}$ -neighourhood of C and $\mathcal{C}_1$ the set of points (x, z) such that $x \in \mathcal{C}$ and $|p_o(x, z)| < \mathcal{E}$. Suppose there exists a function b(x, z) which is homogeneous of degree -1 in z and is the symbol of an operator in $\mathcal{S}_m$ for which

$$\Psi_1 = [p_o, \bar{p}_o] + b \bar{p}_o + p_o \bar{b} \geq 0$$

in $\mathcal{C}_1$. Then , if the P_j are as in 3.3, and .

$$h_1 = p o p^{\#} - p^{\#} \circ p + b^{\#} \circ p + p^{\#} \circ b - b^{\#} \circ b$$

we have

$$P_1^{-1}(h_1) = 0, \quad P_2^{-1}(h_1) = \Psi_1$$

and if

$$P_n^{-1}(h_1) \geq c \, |z|^{-2s}, \quad c > 0$$

for $0 \leq 2s < n = \left[\frac{2}{3} m\right]$ and all (x, z) in $\mathcal{O}_1$ with $|z|$ sufficiently large, the operator $T T^* - c\Lambda^{-2s}$ is almost positive of order n in C for some positive c.

Furthermore, let the symbol $a(x, z)$ of an operator in $\mathcal{S}_m$ be real and homogeneous of degree zero in z. Let $a_1 = \exp(\exp\ a)$ and $p_1 = a_1^{-1} \circ p \circ a_1$, where $a_1^{-1} \circ a_1 = 1$. Then if

$$h_2 = p_1 \circ p_1^{\#} - p_1^{\#} \circ p_1 + b^{\#} \circ p_1 + p_1^{\#} \circ b - b^{\#} \circ b$$

we have

$$P_1^{-1}(h_2) = 0, \quad P_2^{-1}(h_2) = \psi_1, \quad P_3^{-1}(h_2) = P_3^{-1}(h_1) + \psi_2$$

$$\psi_2 = (2\left|[p_o, a]\right|^2 - [\bar{p}_o, [p_o, a]] - [p_o, [\bar{p}_o, a]])\ \exp\ a +$$

$$+ (\bar{b}\,[p_o, a] + b\,[a, p_o])\ \exp\ a.$$

and if

$$P_n^{-1}(h_2) \geq c|z|^{-2s}, \quad c > 0$$

for all (x, z) in $\mathcal{O}_1$ with $|z|$ sufficiently large, the same conclusion as above holds for $T T^* - c\Lambda^{-2s}$.

This is merely a specialization of the last result in 4.1. We take $\sigma(B) = \sigma(T) - b(x, z)$ and $H_1 = T T^* - B^* B$, $h_1 = \sigma(H_1)$ to obtain the first part of our statement. For the second we take $\sigma(A_1) = a_1^{-1}$, $\sigma(A) = a_1$ and $\sigma(B) = \sigma(T_1) - b$. We leave to the reader the verification of details. Now let us illustrate this last statement with a simple example. Let us suppose that the set C is bounded in the direction of x_1 and that p_o is real. Let us take $b = 0$ and $a(x, z)$ so that it coincides with $t(x_1 + c)$ in an ε-neighborhood of C. Then a simple calculation shows that

$$P_2^{-1}(h_2) = 0, \quad \psi_2 = \left(\frac{t^2}{2\pi^2}\left(\frac{\partial p_o}{\partial z_1}\right)^2 + \frac{t}{2\pi^2}\sum \frac{\partial p_o}{\partial z_j}\frac{\partial^2 p_o}{\partial z_1 \partial x_j} - \frac{\partial p_o}{\partial x_j}\frac{\partial^2 p_o}{\partial z_1 \partial z_j}\right) \exp t(x_1 + c)$$

so that if the summation in the last expression is distinct from zero and of constant sign on the set where $\frac{\partial p_o}{\partial z_1} = 0$ taking t and c positive or negative according to the case, and sufficiently large in absolute value, we will have

$$P_3^{-1}(h_2) \geqslant |z|^2 .$$

Clearly, explicit calculations using the formulas of 3.3 will lead to much more precise results. These, however, we leave to the reader .

To conclude, we would like to point out that all results obtained so far extend to matrices of operators. One has only to keep in mind that multiplication of matrices is non-commutative and that the condition for invertibility is the non-vanishing of the determinant. Thus, for example, this last statement is valid for matrices of operators provided that p is multiple of the identity matrix at each point. This condition can be satisfied in all cases by multiplying T on the right by an operator whose principal symbol is the matrix of the complementary minors of p .

4.3 In this paragraph we will discuss some necessary conditions for almost positivity. These conditions are simple consequences of the material presented in section 2 .

Consider the class of functions f_t depending on the parameter t, $t>0$, which are finite sums of the form

$$f_t = f_o + f_1 t^{s_1} + f_2 t^{s_2} + \dots , \quad 0<s_1<s_2<\dots ,$$

where the f_j are functions in $\mathcal{S}$, and let $\mathcal{J}_s$ the subclass of f_t for which $f_j = 0$ for $s_j \leqslant s$.

Let T be an operator in $\mathcal{S}_m$ and suppose that $T\,T^* - c\,\Lambda^{-2s}$, $c>0$, is almost positive of order n, $0\leq 2s<n\leq m$, in an open set $\mathcal{O}$. Let $\lambda(T)$ be the the operator associated with T and the point (x,ν) , $x\in\mathcal{O}$, as in (16). Then for every f_t with $f_o \neq 0$ we have $\lambda(T)^* f_t \notin \mathcal{J}_{2s}$.

Let us show this assuming for simplicity that $x=0$. Let H_t be the operator introduced in section 2 and let φ be an infinitely differentiable function with complact support contained in $\mathcal{O}$ such that $\varphi=1$ in a neighbourhood of x . Then

$$\varphi H_t f_t = H_t f_t + g_t, \quad \| g_t \| = 0(t^{2m}), \quad t \leq 1,$$

the norm here being an L^2-norm. In fact, as readily seen , we have

$$\varphi H_t f_t = H_t \left[f_t(x) \varphi(tx) \right] = H_t f_t + H_t \left[\varphi(tx) - 1 \right] f_t = H_t f_t + g_t$$

Now since H_t is unitary, and assuming that $\varphi(x)=1$ for $|x| < \delta$, we have

$$\| g_t \|^2 = \| \left[\varphi(tx) - 1 \right] f_t \|^2 \leq c \int_{\delta/t < |x|} | f_t(x) |^2 dx = 0(t^{4m})$$

the last estimate being a consequence of the fact that $f_t(x)$ decreases rapidly as $|x| \to \infty$.

Now , on account of the results in section 2 we have

$$(T\, T^* \varphi H_t f_t, \varphi H_t f_t) = (T\, T^* H_t f_t, H_t f_t) + 0(t^{2m}) =$$

$$= \| H_t^* T^* H_t f_t \|^2 + 0(t^{2m}) = \| \sigma(T)^* f_t \|^2 + 0(t^{2m})$$

and similarly

$$(\Lambda^{-2s} \varphi H_t f_t, \varphi H_t f_t) = \| \sigma(\Lambda^{-s}) f_t \|^2 + 0(t^{2m}) = t^{4s} \| f_o \|^2 + 0(t^{4s+\xi}) +$$

$$+ 0(t^{2m})$$

and

$$(\Lambda^{-n} \varphi H_t f_t, \varphi H_t f_t) = t^{2n} \| f_o \|^2 + 0(t^{2n+\xi}) + 0(t^{2m})$$

Now the operator $T\, T^* - c \Lambda^{-2s}$ was assumed to be positive of order n in $\mathcal{O}$. Therefore if $R = T\, T^* - c \Lambda^{-2s} + c_1 \Lambda^{-n}$ and c_1 is positive and sufficiently large, we will have $(R f, f) \geq 0$ for every f with support in $\mathcal{S}$. Thus from the preceding estimates we obtain

$$0 \leq (R\varphi H_t f_t, \varphi H_t f_t) = \|\lambda(T)^* f_t\|^2 - (c\, t^{4s} - c_1 t^{2n}) \|f_o\|^2 + o(t^{4s})$$

whence it follows that for t sufficiently small we will have

$$\|\lambda(T)^* f_t\| \geq c\, t^{2s} \quad , \quad c > 0$$

which clearly implies that $\lambda(T)^* f_t \notin \mathcal{J}_{2s}$, as we wished to show.

Now we will discuss some conditions under which there exists an f_t with $f_o \neq 0$ such that $\|\lambda(T)^* f_t\| = O(t^{2m})$

Let T be an operator in $\mathcal{J}_m$ such that $\sigma(t)$ is a sum of homogeneous functions of negative integral degree in z. Let (x, ν) be a point and

$$\lambda(T) = \lambda_o + \lambda_1 t + \lambda_2 t^2 + \ldots \lambda_{2m-1} t^{2m-1}$$

the operator associated with T and the point (x, ν) as in (16). Suppose that either $\lambda(T) = 0$ or else the first non-vanishing coefficient λ_j has the property that there exists a polynomial $P(x) = P_1(x) + iP_2(x)$ with $P_1(x) < -|x|^{\delta}$, $\delta > 0$, for $|x|$ sufficiently large, and such that the operator

$$\mu_j g = e^{-P} \lambda_j^*(e^P g)$$

has constant coefficients and vanishing constant term. Then there exists f_t

$$f_t = f_o + f_1 t + f_2 t^2 + \ldots$$

with f_r in $\mathcal{S}$ and $f_o \neq 0$, such that $\|\lambda(T)^* f_t\| = 0(t^{2m})$. Thus, for no s and n, $0 \leq 2s < n \leq m$, can the operator $T T^* - c \Lambda^{-2s}$, $c > 0$, be almost positive of order n in a neighbourhood of the point x.

In particular, if j=1, the polynomial P(x) exists provided the constant $\lambda_1 \lambda_1^* - \lambda_1^* \lambda_1$ has negative real part.

This last part is a well known result of Hörmander. To prove our assertion we start observing that the function f_t will satisfy the condition $\|\lambda(T)^* f_t\| = 0(t^{2m})$ if and only if the f_r satisfy the system of equations

$$\lambda_j^* f_o = 0$$

$$\lambda^*_{j+1} f_o + \lambda_j f_1 = 0$$

(28)
$$\lambda^*_{j+2} f_o + \lambda^*_{j+1} f_1 + \lambda^*_j f_2 = 0$$

$$\dots\dots\dots\dots\dots\dots\dots\dots$$

$$\lambda^*_{2m-1} f_o + \lambda^*_{2m-2} f_1 + \dots \lambda_j f_{2m-1-j} = 0 .$$

Now, let us set $f_j = e^P g_j$, substitute above and multiply on the left by e^{-P}, and the equations become

$$\mu_j g_o = 0$$

$$\mu_{j+1} g_o + \mu_j g_1 = 0$$

$$\dots\dots\dots\dots\dots\dots\dots$$

$$\mu_{2m-1} g_o + \mu_{2m-2} g_1 + \dots + \mu_j g_{2m-1-j} = 0$$

where the μ_r are differential operators with polynomial coefficients and μ_j has constant coefficients and vanishing constant term. Evidently, the image of a polynomial under μ_r is again a polynomial, and since, as we shall see below, the equation $\mu_j g = Q$, Q being a polynomial, can always be solved with a polynomial g, setting $g_o = 1$ we can find step by step polynomials g_1, g_2, ... satisfying the equations above. Thus, the system (28) has a solution $f_j = g_j e^P$. Since the g_j are polynomials and the real part of P tends to $-\infty$ like $-|x|^{\varepsilon}$, $\varepsilon > 0$, or faster, the functions f_j belong to $\mathcal{S}$.

Now let us show that every non-vanishing differential operator with constant coefficients maps the space of polynomials onto itself, so that μ_j has indeed the property postulated. Let us consider the space of formal power series Q, and for $Q = \sum b_\alpha x^\alpha$ and the polynomial $P = \sum a_\alpha x^\alpha$ let

$$L(Q,P) = \left[Q\left(\frac{\partial}{\partial x}\right) P\right](0) = \sum b_\alpha a_\alpha \alpha!$$

Evidently, every linear functional on the space of polynomials is of the form L(Q, P) for some Q. Suppose now that R is an operator with constant coefficients whose range as an operator on the space of polynomials is a proper subspace of the latter. Then there exists a non-vanishing Q such that L(Q, RP) = 0 for all polynomials P. But L(QR,P)= = L(Q, RP)= 0 and therefore QR = 0. Since $Q \neq 0$ it follows that R = 0.

Finally, let us turn to the case $\lambda_j = \lambda_1$ and let us show that, under the condition that the real part of $\lambda_1 \lambda_1^* - \lambda_1^* \lambda_1$ be negative, there exists a quadratic form P with negative definite real part such that μ_1 has constant coefficients and vanishing constant term. Let

$$\lambda_1^* = \sum a_j x_j + b_j \frac{\partial}{\partial x_j}$$

Then

$$\mu_1 = \sum a_j x_j + b_j\left(\frac{\partial}{\partial x_j} + \frac{\partial P}{\partial x_j}\right)$$

so that P must satisfy the equation

$$\sum a_j x_j + b_j \frac{\partial P}{\partial x_j} = 0 \tag{29}$$

Suppose now that the real and imaginary parts of the vector $(b_1, b_2, \ldots, b_k)$ are linearly dependent. Then an appropriate change of independent variables will bring the preceding equation to the form

$$b\left[\frac{\partial P}{\partial y_1} P + \sum w_j y_j\right] = 0$$

Since the quantity

$$\sum \bar{b}_j a_j + b_j \bar{a}_j = \lambda_1^* \lambda_1 - \lambda_1 \lambda_1^*$$

is invariant under linear substitutions of variables, w_1 in the equation above must have positive real part, and the quadratic form

$$P = -\frac{1}{2} w_1 y_1^2 - \sum_2^k w_j y_1 y_j - c \sum_2^k y_j^2 .$$

which clearly satisfies the equation, will have negative definite real part for c positive and sufficiently large.

Suppose now that the real and imaginary parts of the vector $(b_1, b_2, \ldots, b_k)$ are linearly independent. Then, again, a linear substitution of independent variables will bring (29) to the form

$$\frac{\partial}{\partial y_1} P + i \frac{\partial}{\partial y_2} P + \sum w_j y_j = 0$$

Setting $w_1 = u_1 + i v_1$, $w_2 = u_2 + i v_2$, on account of the invariance of $\sum \bar{b}_j a_j + b_j \bar{a}_j$ we will have $u_1 + v_2 > 0$, and the quadratic form

$$P = -\frac{1}{4}(u_1+v_2)(y_1^2+y_2^2) - \frac{i}{2}\left[v_1 y_1^2 + (v_2 - u_1) y_1 y_2 - u_2 y_2^2\right] - \sum_3^k w_j y_1 y_j - c \sum_3^k y_j^2$$

which clearly satisfies the preceding equation, will have negative definite real part for c positive and sufficiently large. Thus our assertion is established.

To conclude, we would like to point out that the same method can be employed to obtain conditions in order that μ_j have the required properties. For example, for a given j one might establish the conditions for a suitable μ_j to be obtainable with P a quadratic form. This leads to the problem of solvability in the real domain of algebraic equations and inequalities which can be explicitly computed, and which, in turn, is equivalent with certain algebraic inequalities being satisfied by the coefficients of λ_j.

REFERENCES

1 Calderón A.P. "Algebras of singular integral operators" Proc. Symposia Pure Math., A.M.S., vol. 10, pp. 18-55.

2 .... "Estimate for singular integral operators", to appear.

3 .... "Lebesgue spaces of differentiable functions and distributions" Proc. Symposia Pure Math., A.M.S., vol. 4

4 Calderón A.P. and Zygmund A., "Singular integral operators and differential equations", Amer. J. Math. 79, (1957), 901-921

5 Hörmander L., "Pseudo-differential operators and non-elliptic boundary problems", Ann. of Math., 83, (1966), 129-209.

6 Kohn J.J. and Nirenberg L. "An algebra of pseudo-differential operators" Comm. Pure Appl. Math. 18 (1965), 269-305.

7 Mihlin S.G., " On the multipliers of Fourier integrals", Dokl. Akad. Nauk SSSR, 109, (1956), 701-703

CENTRO INTERNAZIONALE MATEMATICO ESTIVO

(C.I.M.E.)

B.F. JONES

CHARACTERIZATION OF SPACES OF BESSEL POTENTIALS RELATED TO THE HEAT EQUATION

Corso tenuto a Stresa dal 26 agosto al 3 settembre 1968

CHARACTERIZATION OF SPACES OF BESSEL POTENTIALS RELATED TO THE HEAT EQUATION

by

B. Frank JONES Jr.

(Rice University)

1. Background. The operator Λ^{ω} is often dealt with in the talks in this conference. For complex ω it is defined by the formula in the Fourier transform space,

$$\widehat{\Lambda^{\omega} f}(\xi) = (1+|\xi|^2)^{\omega/2}\hat{f}(\xi),$$

where $|\xi|$ is the Euclidean length of ξ. Thus, if Δ is the Laplace operator, it follows that formally

$$\Lambda^{\omega} = (1-\Delta)^{\omega/2} .$$

This gives an obvious and useful relation between Λ^{ω} and Δ. Consider now the heat operator $\Delta - \partial/\partial t$ on R^{n+1}. If one writes points in R^{n+1} as (x,t), (y,s), etc., with x, y, etc. $\in R^n$, and points in the dual R^{n+1} as (ξ,τ), with $\xi \in R^n$, then

Definition 1. For any $\omega \in C$ the operator $\mathcal{J}^{\omega}$ is given by

$$\widehat{\mathcal{J}^{\omega} f}(\xi,\tau) = (1+|\xi|^2+i\tau)^{-\omega/2}\hat{f}(\xi,\tau).$$

If $\mathcal{S}$ is the usual space of C^{∞} functions on R^{n+1} all of whose derivatives decay at ∞ more rapidly than any rational function, and $\mathcal{S}'$ is its dual space, then it follows immediately that

$$\mathcal{J}^{\omega} : \mathcal{S} \to \mathcal{S},$$
$$\mathcal{J}^{\omega} : \mathcal{S}' \to \mathcal{S}'.$$

Definition 2. For $\alpha \in R$ and $p \in [1,\infty]$, $\mathcal{L}^p_{\alpha}$ is the space $\mathcal{J}^{\alpha}(L^p(R^{n+1}))$. For $T \in \mathcal{L}^p_{\alpha}$, define

$$\|T\|_{\mathcal{L}^p_{\alpha}} = \|\mathcal{J}^{-\alpha} T\|_{L^p} .$$

It is easily checked that $\mathcal{L}^p_\alpha$ is a Banach space, and that $\mathcal{G}^\omega$ is an isometry of $\mathcal{L}^p_\alpha$ onto $\mathcal{L}^p_{\alpha+\omega}$.

The purpose of this talk is to describe the spaces $\mathcal{L}^p_\alpha$ for $\alpha>0$ completely in terms of properties of functions in L^p, without recourse to the Fourier transform. The spaces $\mathcal{L}^p_\alpha$ can be termed spaces of potentials because if

$$\widehat{\mathcal{G}^\alpha}(\xi,\tau) = (1+|\xi|^2+i\tau)^{-\alpha/2},$$

then $\mathcal{G}^\alpha \in L^1$, and $T\in\mathcal{L}^p_\alpha \Leftrightarrow \exists\, \varphi\in L^p$ such that $T=\mathcal{G}^\alpha * \varphi$ [4]. Note then that $\mathcal{L}^p_\alpha \subset L^p$. They are termed Bessel potentials because of the presence of the 1 in the definition of $\widehat{\mathcal{G}^\alpha}$. The other kind of potential would be called a Riesz potential, the kernel $\mathcal{H}^\alpha$ having the definition

$$\widehat{\mathcal{H}^\alpha}(\xi,\tau) = (|\xi|^2+i\tau)^{-\alpha/2}.$$

This makes sense as long as $\alpha<n+2$, so that the right side is in L^1_{loc}. For $0<\alpha<n+2$, also $\mathcal{H}^\alpha$ is in L^1_{loc} and, explicitly,

(1) $$\mathcal{H}^\alpha(x,t) = \begin{cases} (4\pi)^{-n/2}\Gamma(\alpha/2)^{-1}\, t^{\frac{\alpha-n}{2}-1}\exp(-|x|^2/4t), & t>0,\\ 0, & t<0.\end{cases}$$

This formula is the one which shall be taken as the definition of $\mathcal{H}^\alpha$ for all $\alpha\in R$; the formula for $\widehat{\mathcal{H}^\alpha}$ then holds for $0<\alpha<n+2$.

The connection of this talk to pseudo-differential operators is entirely in the use of the estimates belonging to Calderón-Zygmund singular integrals, which are special types of pseudo-differential operators: The estimates needed here will always be for singular integrals with mixed homogeneity. One of the prime examples of this is in obtaining a sufficient condition for a function to be a multiplier in the Mihlin sense. Such a sufficient condition can be found in e.g. [3], [5]. One use of a result like this is in the following (see [1])

Proposition 1. *If* k *is a non-negative integer and* $\alpha\geq 2k$ *and*

$p \in (1, \infty)$, then

$$f \in \mathcal{L}^p_\alpha \Leftrightarrow D_x^\beta D_t^\gamma f \in \mathcal{L}^p_{\alpha-2k}, \quad |\beta| + 2\gamma \leq \alpha.$$

Remarks. (a) There is also a norm equivalence here; namely, $\|f\|_{\mathcal{L}^p_\alpha}$ and

$$\sum_{|\beta|+2\gamma \leq \alpha} \|D_x^\beta D_t^\gamma f\|_{\mathcal{L}^p_{\alpha-2k}}$$

are equivalent norms in the normed space sense. Such equivalence will hold throughout this article and will not usually be explicitly mentioned.

(b) As is usual in using results obtained via singular integrals techniques, the cases $p=1$ and $p=\infty$ are not included.

(c) This proposition shows that for $p \in (1, \infty)$, the explicit characterization of $\mathcal{L}^p_\alpha$ which is desired need be obtained only for $0 < \alpha < 2$.

This ends the background material, except for the important acknowledgments to my students, Richard Bagby and Charles Sampson, who kindly consented to my describing some results of their unpublished theses in this talk. The new theorems here are theirs.

2. Sampson's characterization. Following E.M. Stein [8], Sampson [7] has obtained a characterization of $\mathcal{L}^p_\alpha$ for $0 < \alpha < \infty$, $1 \leq p \leq \infty$. To describe this, it is necessary to introduce the difference operator $\Delta^{(2m)}_{y,s}$, defined for a positive integer m by

Definition 3. For a function f on R^{n+1}, and $(z, s) \in R^{n+1}$,

$$(\Delta^{(2m)}_{y,s} f)(x,t) = \sum_{j=0}^{2m} \binom{2m}{j} (-1)^j f(x-(m-j)y,\ t-(m-j)^2 s).$$

Now Sampson gives the following results. First, if $\alpha > 0$ then the definition

$$\widehat{(-\Delta + \partial/\partial t)^{\alpha/2} \varphi} = (|\xi|^2 + i\tau)^{\alpha/2} \hat{\varphi}$$

makes sense for $\varphi \in \mathcal{S}$, and clearly

$$(-\Delta + \partial/\partial t)^{\alpha/2} : \mathcal{S} \to L^\infty .$$

Much more is true : if $p \in [1, \infty]$, this operator maps $\mathcal{S}$ into L^p and has a continuous extension

$$(-\Delta + \partial/\partial t)^{\alpha/2} : \mathcal{L}^p_\alpha \to L^p .$$

This extension can be explicitly computed, and the result is :

Theorem 1. Let $p \in [1, \infty]$, $0 < \alpha < 2m$, and let $f \in L^p$. Then $f \in \mathcal{L}^p_\alpha \Leftrightarrow$ (a) if $p < \infty$, the limit

$$(2) \qquad \lim_{\varepsilon \to 0} \int_{s > \varepsilon} \Delta^{(2m)}_{y,s} f(x,t)\, c(\alpha) \mathcal{H}^{-\alpha}(z,s)\,dz\,ds$$

exists in L^p ; (b) if $p = \infty$, the integral in (2) is uniformly bounded, in which case the limit in (2) exists in the weak* topology of L^∞. Here

$$c(\alpha) = 2 \left[\sum_{j=0}^{m-1} \binom{2m}{j} (-1)^j (m-j)^\alpha \right]^{-1} ,$$

and $\mathcal{H}^{-\alpha}$ is the function in (1) with α replaced by $-\alpha$.

Remarks (a) If $\alpha = 2, 4, \ldots, 2m-2$, then $\Gamma(-\alpha/2)^{-1} = 0$ and c has a simple pole at α ; the product $c(\alpha)\mathcal{H}^{-\alpha}$ is still meaningful, however.

(b) The limit in (2) is exactly $(-\Delta + \partial/\partial t)^{\alpha/2} f$.

The result of Theorem 1 is useful in obtaining counter examples to certain inclusion relations which might hold between $\mathcal{L}^p_\alpha$ and Lipschitz spaces [4]. The reasonableness of the difference operator $\Delta^{(2m)}$ is seen from Theorem 1, and also from the fact that $\Delta^{(2m)}$ eliminates from the Taylor's series for f approximately twice as many terms in x as in t, a natural ratio in dealing with the heat equation.

The main disadvantage in the characterization of Theorem 1 is that the difference $\Delta^{(2m)}$ appears without absolute value sign, giving an integral which is somewhat difficult to work with. The next results will not have this disadvantage, but will require $p \in (1, \infty)$.

3. Bagby's characterization. Throughout this section p is fixed, $p \in (1, \infty)$. Also it is sufficient to allow $\alpha \in (0, 2)$, by Remark (c) following Proposition 1. Thus, in considering the difference operator introduced in section 2, it suffices to take $m = 1$, and then

$$(\Delta^{(2)}_{y,s} f)(x,t) = f(x-y, t-s) - 2f(x,t) + f(x+y, t-s).$$

(Note that in section 2 for $p=1$ or ∞, a characterization for all $\alpha > 0$ does not follow from a characterization for $0 < \alpha < 2$.) What will be used in this section is the absolute value $|\Delta^{(2)}_{y,s} f|$.

Remark. If $0<\alpha<1$, then every thing in this section is valid with

$\Delta^{(2)}_{y,s} f$ replaced by

$$f(x-y, t-s) - f(x,t).$$

This is also advantageous, as first differences are generally easier to deal with than second differences. Moreover, since the methods of interpolation are available for the spaces $\mathcal{L}^p_\alpha$ (cf. [1]), it is frequently sufficient to have known results for large even integers and small positive α.

The following results are analogous to results of Strichartz [9], who originally obtained theorems like these for the ordinary Bessel potentials.

Definition 4. Let $\Omega = \{(y,s) \in R^{n+1} : |y| \leq 1, |s| \leq 1$. Define

$$S_\alpha f(x,t) = \left\{ \int_0^\infty r^{-2\alpha} \left[\int_\Omega |\Delta^{(2)}_{ry, r^2 s} f(x,t)| \, dy \, ds \right]^2 dr/r \right\}^{1/2}.$$

Theorem 2. Let $f \in L^p$. Then $f \in \mathcal{L}^p_\alpha \Leftrightarrow S_\alpha f \in L^p$.

As an application the utility of Theorem 2, suppose $1 > \alpha > (n+2)/p$. Then $\mathcal{L}^p_\alpha$ is an algebra under pointwise multiplication. To see this, let $f, g \in \mathcal{L}^p_\alpha$.

Sobolev's inequality (cf. [1]) shows that $\mathcal{L}_\alpha^p \subset L^\infty$. Thus, $fg \in L^p$. And $S_\alpha(fg) \leq \|f\|_\infty S_\alpha g + \|g\|_\infty S_\alpha f \in L^p$, showing $fg \in \mathcal{L}_\alpha^p$. Actually, if $\infty > \alpha > (n+2)/p$, it still is true that $\mathcal{L}_\alpha^p$ is an algebra, as follows from interpolation [1].

The rest of this paper gives an outline of the proof of Theorem 2, showing how it is based on singular integrals. Also, some of the simple computations will be given. Two facts from singular integrals will be needed:

Lemma 1. [2, 6] Let X be a Banach space, $k(x,t) \in X$ for a.e. (x,t). Let A map L^∞_{com} into measurable X-valued functions by the formula

$$(A\varphi)(x,t) = \int \varphi(z,u)\, k(x-z, t-u)\, dz du .$$

Assume

(a) $$\|A\varphi\|_{L^2(X)} \leq c\|\varphi\|_2 \quad \forall \varphi;$$

(b) $$\int_{|x| \geq 2r \text{ or } |t| \geq 4r^2} \|k(x-z,t-u)-k(x,t)\|_X \, dxdt \leq c$$

if $|z| \leq r$ and $|u| \leq r^2$.

Then

$$\|A\varphi\|_{L^p(X)} \leq C_p \|\varphi\|_p , \qquad 1 < p < \infty .$$

(Here $L^p(X)$ is the space of functions on R^{n+1} whose X-norms are in L^p).

Lemma 2. [1, 2] Let H be a Hilbert space and let B be a linear operator such that

$$B : L^p \to L^p(H) , \quad 1 < p < \infty ,$$

and such that there exists an H-valued function R^{n+1} such that

$$\widehat{B\varphi} = \hat{\varphi} h,$$

and suppose

(a) h is bounded,

(b) $\{h(\rho\xi, \rho^2\tau)\}_{0<\rho<\infty}$ is an equicontinuous family near the set $|\xi| + |\tau|^{1/2} = 1$,

(c) $\|B\varphi\|_{L^2(H)} \geq c\,\|\varphi\|_2$, some $c>0$, all $\varphi \in L^2$.

Then $\exists\ c_p > 0$ such that

$$\|B\varphi\|_{L^p(H)} \geq c_p \|\varphi\|_p,\ \varphi \in L^p.$$

Application of Lemma 1. Take X to be the space of functions g on $(0,\infty)\times\Omega$ with

$$\|g\|_X = \left\{\int_0^\infty r^{-2\alpha}\left[\int_\Omega |g(r,y,s)|\,dyds\right]^2 dr/r\right\}^{1/2} < \infty.$$

Let

$$k(x,t)(r,y,s) = \Delta^{(2)}_{ry,\,r^2s}\,\mathcal{H}^\alpha.$$

Then note that

$$(A\varphi)(x,t) = \Delta^{(2)}_{ry,\,r^2s}(\mathcal{H}^\alpha * \varphi)(x,t),$$

so that

$$\|A\varphi(x,t)\|_X = S_\alpha(\mathcal{H}^\alpha * \varphi)(x,t).$$

Thus, if (a) and (b) are satisfied, it follows from Lemma 1 that

$$\|S_\alpha(\mathcal{H}^\alpha * \varphi)\|_p \leq C_p \|\varphi\|_p\,. \tag{3}$$

The verification of (a) is quite simple. In fact, if $g = \mathcal{H}^\alpha * \varphi$, then the Schwarz inequality gives (using M for a generic constant)

$$\|A\varphi(x,t)\|_X^2 \leq M\int_0^\infty r^{-2\alpha}\int_\Omega |\Delta^{(2)}_{ry,\,r^2s}\,g(x,t)|^2\,dyds\ dr/r$$

$$= M\int_0^\infty r^{-2\alpha-n-3}dr\int_{|y|\leq r,\ |s|\leq r^2} |\Delta^{(2)}_{y,s}\,g(x,t)|^2\,dyds$$

$$= M\int_{R^{n+1}} |\Delta^{(2)}_{y,s}\,g(x,t)|^2\,dyds\int_{\max(|y|,\,|s|^{1/2})}^\infty r^{-2\alpha-n-3}dr$$

$$\leqq M \int_{R^{n+1}} (|y| + |s|^{1/2})^{-2\alpha-n-2} |\Delta^{(2)}_{y,s} g(x,t)|^2 \, dy\, ds .$$

By Fubini's theorem and Parseval's relation,

$$\|A\varphi\|^2_{L^2(X)} \leqq M \int (|y| + |s|^{1/2})^{-2\alpha-n-2} dyds \int |\Delta^{(2)}_{y,s} g(x,t)|^2 dxdt$$

$$= M \int (|y|+|s|^{1/2})^{-2\alpha-n-2} dyds \int |\hat{g}(\xi,\tau)|^2 |e^{-iy\cdot\xi-is\tau} + e^{iy\cdot\xi-is\tau} - 2|^2 \, d\xi\, d\tau$$

$$= M \int (|y| + |s|^{1/2})^{-2\alpha-n-2} dyds \; ||\xi|^2 + i\tau|^{-\alpha} |\hat{\varphi}(\xi,\tau)|^2 \, |e^{-iy\cdot\xi-is\tau} + e^{iy\cdot\xi-is\tau} - 2|^2 d\xi\, d\tau .$$

Thus, Parseval's relation again shows that one must prove that

$$||\xi|^2 + i\tau|^{-\alpha} \int (|y| + |s|^{1/2})^{-2\alpha-n-2} |e^{-is\tau}\cos(y\cdot\xi) - 1|^2 dyds$$

is bounded. A change of variables shows this whole expression depends only on $\dfrac{\xi}{||\xi|^2+i\tau|^{1/2}}$ and $\dfrac{\tau}{|\xi|^2+i\tau}$, and thus it suffices to consider $||\xi|^2+i\tau|=1$.

But then

$$|e^{-is\tau}\cos(y\cdot\xi) - 1| \leqq \begin{cases} 2 \\ |y|^2 + |s| , \end{cases}$$

and then the resulting integral is seen to converge since $\alpha < 2$.

The verification of (b) is involved with extremely tedious estimations, using only elementary techniques, and will not be repeated here. See [1] for the complete details.

Application of Lemma 2. Take H to be $L^2(0,\infty)$ with respect to the measure $r^{-2\alpha-1}\,dr$. Define

$$k_r(x,t) = \int_\Omega \Delta^{(2)}_{ry,\, r^2 s} \mathcal{H}^\alpha(x,t) dy\, ds$$

and

$$B\varphi(x,t)(r) = (k_r * \varphi)(x,t) .$$

Then

$$\widehat{B\varphi}(\xi,\tau)(r) = \hat{k}_r(\xi,\tau)\hat{\varphi}(\xi,\tau) ,$$

so that in the notation of Lemma 2

$$h(\xi,\tau)(r) = \hat{k}_r(\xi,\tau).$$

Suppose that Lemma 2 can be applied in this situation. Then

$$B\varphi(x,t)(r) = \int_\Omega \Delta^{(2)}_{ry,\,r^2s} \mathcal{H}^\alpha * \varphi\,(x,t)\,dy\,ds,$$

so that by Definition 4

$$\|B\varphi(x,t)\|_H = \left\{\int_0^\infty r^{-2\alpha} \left| \int_\Omega \Delta^{(2)}_{ry,\,r^2s} \mathcal{H}^\alpha * \varphi(x,t)\,dy\,ds \right|^2 dr/r\right\}^{1/2}$$

(4) $$\leq \quad S_\alpha(\mathcal{H}^\alpha * \varphi)(x,t)\,.$$

The conclusion of Lemma 2 gives therefore

(5) $$\| S_\alpha(\mathcal{H}^\alpha * \varphi)\|_p \geq C'_p \|\varphi\|_p\,,$$

a result which combines very nicely with (3).

Now the hypothesis of Lemma 2 must be checked for this case. First, the inequality (4) shows in conjunction with (3) that $B\colon L^p \to L^p(H)$.

Next,

$$\hat{k}_r(\xi,\tau) = \int_\Omega \widehat{\mathcal{H}^\alpha}(\xi,\tau)\left[e^{-iry\cdot\xi - ir^2 s\tau} + e^{iry\cdot\xi - ir^2 s\tau} - 2\right] dy\,ds$$

so that

(6) $$\|h(\xi,\tau)\|_H^2 = 4\left||\xi|^2 + i\tau\right|^{-\alpha} \int_0^\infty r^{-2\alpha} \left|\int_\Omega \left[\cos(ry\cdot\xi)e^{-ir^2 s\tau} - 1\right] dy\,ds\right|^2 dr/r.$$

A change of variables then shows that

$$\|h(\rho\xi,\rho^2\tau)\|_H = \|h(\xi,\tau)\|_H\,, \quad 0<\rho<\infty\,,$$

so that (a) and (b) are rather easy consequences. For example, if $\left||\xi|^2 + i\tau\right| = 1$, then the integral over Ω in is $O(1)$ and also $O(r^2)$, showing that $\|h(\xi,\tau)\|_H$ is bounded since $0<\alpha<2$. A similar argument gives (b). Finally, Parseval's relation shows that (c) is equivalent to

$$\|h(\xi,\tau)\|_H > 0\,, \quad (\xi,\tau) \neq 0\,.$$

This is easily checked, since the integral over Ω in (6) is approximately the negative of the volume of Ω_ρ for $r \to \infty$.

<u>Proof of Theorem 2</u>. One final lemma is needed:

<u>Lemma 3</u>. [7] <u>If</u> $0<\alpha$, <u>then there exist totally finite Borel measures</u>

μ_1, μ_2, μ_3 on R^{n+1} such that

$$\text{(a)} \quad (|\xi|^2+i\tau)^{\alpha/2} = (1+|\xi|^2+i\tau)^{\alpha/2}\hat{\mu}_1 \;;$$

$$\text{(b)} \quad (1+|\xi|^2+i\tau)^{\alpha/2} = \hat{\mu}_2 + (|\xi|^2+i\tau)^{\alpha/2}\hat{\mu}_3.$$

Now let $\varphi \in \mathcal{S}$ and let $f = \mathcal{J}^\alpha\varphi$. Then

$$\hat{f} = (1+|\xi|^2+i\tau)^{-\alpha/2}\hat{\varphi}$$

$$= (|\xi|^2+i\tau)^{-\alpha/2}\hat{\mu}_1\hat{\varphi},$$

so that

$$f = \mathcal{H}^\alpha * \mu_1 * \varphi.$$

Therefore, (3) implies

$$\| S_\alpha f\|_p \leq C_p\|\mu_1 * \varphi\|_p \leq C_p\|\mu_1\| \|\varphi\|_p = C_p\|\mu_1\| \|f\|_{\mathcal{L}^p_\alpha} .$$

Conversely, Lemma 3(b) implies

$$\hat{\varphi} = (1+|\xi|^2+i\tau)^{-\alpha/2}\hat{\mu}_2\hat{\varphi} + (1+|\xi|^2+i\tau)^{-\alpha/2}(|\xi|^2+i\tau)^{\alpha/2}\hat{\mu}_3\hat{\varphi}.$$

$$= \hat{\mu}_2\hat{f} + \hat{\mu}_3(|\xi|^2+i\tau)^{\alpha/2}\hat{f},$$

so that

$$\varphi = \mu_2 * f + \mu_3 * (-\Delta + \partial/\partial t)^{\alpha/2} f \;.$$

Therefore, (5) implies

$$\|f\|_{\mathcal{L}^p_\alpha} = \|\varphi\|_p \leq \|\mu_2\|\|f\|_p + \|\mu_3\|\|(-\Delta+\partial/\partial t)^{\alpha/2} f\|_p$$

$$\leq M\|f\|_p + M\|S_\alpha(\mathcal{H}^\alpha * (-\Delta+\partial/\partial t)^{\alpha/2} f)\|_p$$

$$= M\|f\|_p + M\|S_\alpha f\|_p \;.$$

This finishes the proof except for the details involved in the various convolution and Fourier transform formulas above, and i making the extension to $\varphi \in L^p$.

References

[1] R. Bagby, Lebesgue spaces of parabolic potentials, Thesis, Rice Univ., 1968

[2] A. Benedek, A. Calderón, and R. Panzone, Convolution operators on Banach space valued functions, Proc. Nat. Acad. U.S.A. 48(1962), 356-365.

[3] E. Fabes and N. Rivière, Singular integrals with mixed homogeneity, Studia Math. 27(1966) 19-38.

[4] F. Jones, Lipschitz spaces and the heat equation, to appear in J. Math. Mech.

[5] P. Krée, sur les multiplicateurs dans L^p, C.R. Acad. Sci. Paris 260 1965, 4400-4403.

[6] J. Lewis, Mixed estimates for singular integrals and an application to initial value problems of parabolic type, Symposium on singular integrals, Chicago, 1966.

[7] C. Sampson, A characterization of parabolic Lebesgue spaces, Thesis, Rice Univ., 1968.

[8] E. Stein, The characterization of functions arising as potentials, Bull. Amer. Math. Soc. 67(1961) 102-104.

[9] R. Strichartz, Multipliers on fractional Sobolev spaces, J. Math. Mech. 16(1967) 1031-1060.

CENTRO INTERNAZIONALE MATEMATICO ESTIVO

(C. I. M. E.)

J. J. KOHN

PSEUDO-DIFFERENTIAL OPERATORS AND NON-ELLIPTIC PROBLEMS

Corso tenuto a Stresa dal 26 Agosto al 3 Settembre 1968

PSEUDO - DIFFERENTIAL OPERATORS AND NON-ELLIPTIC PROBLEMS

by

J. J. KOHN [1)]

We will discuss here problems which lead to integro-differential forms which involve derivatives of first order. Such a problem is called elliptic when the L_2-norms of all first derivatives can be estimated by the corresponding form. In [4] it is shown that it suffices to estimate the $\| \|_\varepsilon$-norm (with $0 < \varepsilon \leq 1$) in order to establish smoothness of solutions, discretness of spectrum and other properties such problems have in common with the elliptic case. Here we consider the case when the L_2-norms of only some derivatives are bounded.

The following is a special case of an estimate proved by Hörmander in 1 . The proof that we give here uses only elementary properties of pseudo--differential operators.

Let U be a bounded open set in $\mathbb{R}^n$ and let $X_1, \ldots, X_k$ be vector fields in a neighborhood of $\overline{U}$, we have

$$(1) \qquad X_j = \sum_i a_j^i \frac{\partial}{\partial x_i}, \qquad j = 1, \ldots, k,$$

where, the a_j^i are C^∞ functions on a neighborhood of $\overline{U}$.

<u>Theorem</u>. If there is a neighborhood of $\overline{U}$ on which the Lie algebra generated by $X_1, \ldots, X_k$ equals the Lie algebra of all vector fields, then there exists $\varepsilon > 0$ and $C > 0$, such that :

$$(2) \qquad \|u\|_\varepsilon \leq C\left(\sum_{i=1}^{k} \|X_i u\| + \|u\|\right),$$

for all $u \in C_0^\infty(U)$.

The proof uses the following elementary properties of pseudo-differential operators, which have been discussed in Seeley's lecture and can also

1) During the preparation of this paper the author was supported, in part, by a research project sponsored by the N.S.F. at Brandeis University. Part of this work was also prepared under a project ARO. O.

be found in [5] .

(A) A pseudo-differential operator T is of order s if for each real number r there exist a constant C_r such that :

3) $$\| T u \|_r \leq C_r \| u \|_{s+r} ,$$

for all $u \in C_0^\infty$.

(B) If T, T' are pseudo-differential operators of orders s, s' respectively; then T^*, T+T' , T T' and T, T' are pseudo differential operators of orders s, max (s, s') , s+s' and s+s'-1 respectively. Here T^* denotes the L_2-adjoint of T, T T' is the composition of T and T' and $[T, T'] = T T' - T' T$ is the commutator of T and T' .

(C) The operator Λ^s , defined by

4) $$(\Lambda^s u)^\wedge (\xi) = (1+|\xi|^2)^{\frac{s}{2}} \hat{u} (\xi) ,$$

is pseudo-differential operator of order s , where s in any real number. If s is a non negative integer then a linear differential operator with C^∞ coefficients of order s is a pseudo-differential operator of order s .

(D) $\|u\|_s = \| \Lambda^s u \|$ and there exist positive constants C and C' such that,

5) $$C \sum_{i=1}^{n} \left\| \frac{\partial u}{\partial x_i} \right\|_{s-1} \leq \| u \|_s \leq C' \sum_{i=1}^{u} \left\| \frac{\partial u}{\partial x_i} \right\|_{s-1}$$

for all $u \in C_0^\infty$.

Before proving the theorem we state the lemma which is easily proved using the Jacobi identity and induction.

<u>Lemma.</u> The Lie algebra generated by $X_1, \ldots, X_k$ is spanned by elements of the form

6) $$\left[X_{i_1}, \left[X_{i_2}, \ldots, \left[X_{i_{p-1}}, X_{i_p}\right] \cdots \right]\right] \quad ;$$

i.e. each element of the Lie algebra can be expressed as a linear combination with C^∞ coefficients of elements in the above form.

Proof of the theorem: We wish to estimate $\|u\|_\varepsilon$ with ε to be determined later. By property (D) it suffices to estimate the $\|\frac{\partial u}{\partial x_i}\|_{\varepsilon-1}$. By the assumption of the theorem and by the lemma, we can write:

$$\frac{\partial u}{\partial x_i} = \sum_j c_i^j \, F_{pj} \, u \ ,$$

where the F_{pj} are elements of the form (6) involving p_j of the X's and c_i^j are C^∞ in a neighborhood of U. Thus it suffices to estimate the $\| c_i^j F_{pj} u \|_{\varepsilon-1}$ and since c_i^j is pseudo-differential operator of order zero it suffices, by virtue of (A), to estimate $\| F_{pj} u \|_{\varepsilon-1}$. Now, dropping subscripts, we write

$$F_p = \left[X, F_{p-1}\right]$$

and we have :

7) $$\|F_p u\|^2_{\varepsilon-1} = (\Lambda^{\varepsilon-1} F_p u, \Lambda^{\varepsilon-1} F_p u) = (T^{2\varepsilon-1} F_p u, u)$$
$$= (T^{2\varepsilon-1} X F_{p-1} u, u) - (T^{2\varepsilon-1} F_{p-1} X u, u) \ ,$$

where $T^{2\varepsilon-1} = F_p^* \Lambda^{2\varepsilon-2}$ is an pseudo-differential operator of order $2\varepsilon-1$. We will denote by T^s pseudo-differential operators of order s. The first term on the right side of (7) can be written as follows:

$$(T^{2\varepsilon-1} X F_{p-1} u, u) = (\left[T^{2\varepsilon-1}, X\right] F_{p-1} u, u) +$$
$$+ (T^{2\varepsilon-1} F_{q-1} u, X^* u)$$

since $[T^{2\varepsilon-1}, X]$ is of order $2\varepsilon-1$ and since $X^* = -X + T^0$ we have the following estimate

8) $|(T^{2\varepsilon-1} X F_{p-1} u, u)| \lesssim \text{const} (\| F_{p-1} u \|^2_{2\varepsilon-1} + \| Xu \|^2 + \| u \|^2)$.

Similarly the second term on the right hand side of (7) can be written as:

$$(T^{2\varepsilon-1} F_{p-1} Xu, u) = (Xu, ([T^{2\varepsilon-1}, F_{p-1}])^* u) + (Xu, T^{2\varepsilon-1} F^*_{p-1} u)$$

and since $([T^{2\varepsilon-1}, F_{p-1}])^*$ is of order $2\varepsilon-1$ and $F^*_{p-1} = -F_{p-1} + T^0$ we have the estimate

9) $|(T^{2\varepsilon-1} F_{p-1} Xu, u)| \lesssim \text{const} (\| F_{p-1} u \|^2 + \| Xu \|^2 + \| u \|^2_{2\varepsilon-1})$.

Assuming that $2\varepsilon-1 \leq 0$, we have $\| u \|_{2\varepsilon-1} \lesssim \text{const} \| u \|$, and combining this with (8), (9) and (7) we have

(10) $$\| F_p u_{\varepsilon-1} \| \lesssim \text{const.} (\| F_{p-1} \|_{2\varepsilon-1} + \| Xu \| + \| u \|).$$

Replacing p by p-1 and ε by 2ε we can obtain a similar estimate for $\| F_{p-1} \|_{2\varepsilon-1}$ provided $4\varepsilon-1 \leq 0$. Repeating p-times we obtain

(11) $$\| F_p u \|_{\varepsilon-1} \lesssim \text{const} \left(\sum_{i=1}^{k} \| X_i u \| + \| u \| \right),$$

provided that $2^p \varepsilon - 1 \leq 0$. Now choosing p to be the largest of the p_j occuring in the expressions for the $\frac{\partial}{\partial x_j}$ we obtain the desired inequality (2) for $\varepsilon \lesssim 2^{-p}$. Which proves the theorem.

.-.-.-.

As an example of a problem where the type of estimate discussed above consider the system A: E → F, where E consists of p-tuples of C^∞ functions F of q-tuples of C^∞ functions and A is the first order operator given by :

12) $$(A\,u)^j = \sum_{|\varepsilon|<1} \sum_i a^j{}_i\, D\, u^i \,, \qquad j = 1,\ldots,q \,.$$

We are intersted in the system

13) $$A u = f \,,$$

in particular, given f when does there exist u satisfying (13) and if it exists how does it depend on f.? We must consider the so-called compatibility conditions, i.e. if B is an operator such that $BA = 0$ then obviously f must satisfy $B f = 0$. In [4] we show that the following estimate yields the desired results :

14) $$\|v\|_{\varepsilon} \leq \text{const.}\,(\|A^* v\| + \|B v\| + \|v\|) \,,$$

assuming that B is of first order. Again the ellyptic case corresponds to $\varepsilon=1$ and has been discussed by Singer in his lectures. In certain cases the operators A and B can be written in terms of vector fields $X_1,\ldots,X_k$ and the expression $\|X_i v\|$ can be dominated by the right side of (14). To see how this situation can wrise naturally, let M be a real (2n-1)-dimensional compact manifold in $\mathbb{C}^n$. If the complex coordinates on $\mathbb{C}^n$ are $z_1,\ldots,z_n$ with $x_j=\mathrm{Re}\, z_j$, $y_j=\mathrm{Im} z_j$ we consider the inhomogeneous Cauchy-Riemann equations

(15) $$\frac{\partial u}{\partial \bar{z}_j} = \frac{1}{2}\left(\frac{\partial u}{\partial x_j} + \sqrt{-1}\,\frac{\partial u}{\partial y_j}\right) = f_j$$

and the corresponding compatibility conditions

(16) $$\frac{\partial f_j}{\partial \bar{z}_h} - \frac{\partial f_h}{\partial z_j} = 0 \,.$$

Now this complex is elliptic. However, if we restrict it to M, by considering those combinations of the operators (15) and (16) which are tangential to M, then we obtain differential complex that is expressed in terms of 2n-2 vector fields on a (2n-1)-dimensional manifold. This complex is not elliptic, but under certain circumstancys the estimate

(14) holds (with $\varepsilon = \frac{1}{2}$). This example is studied in detail in [2] and [3].

.-.-.- -.-.-.-.-.-.

In conclusion we want to mention the theorem of Hörmander (see 1). Let P be a second order operator, given by:

$$P u = \sum_{i=1}^{k} X_i^2 u + X_o u + c u , \tag{17}$$

where $X_o, \ldots, X_k$ are vector fields and c is in C^∞. It is then easy to see that ;

$$\sum_{i=1}^{k} \| X_i u \|^2 \leq \text{const.} \, (-(Pu, u) + \| u \|^2) . \tag{18}$$

Hörmander proves that P is hypoellyptic if the Lie algebra generated by $X_o, \ldots, X_k$ is the Lie algebra of all vector fields. Our estimate here, combined with (18), will only give the result if the Lie algebra of all vector fields is generated by $X_1, \ldots, X_k$. Pushing the methods used here a little further we comprove the result if the Lie algebra of all vector fields is generated by $X_1, \ldots, X_k$ and elments containing X_o at most three times. However we expect that our mehtod will yield Hörmander's result if we are able to generalize the notion of pseudo-differential operators to include "approximate roots" of vector fields.

J. J. Kohn

REFERENCES

1 HÖRMANDER, L. ," Hypoelliptic second order differential equations" Acta Math. Vol. 119 (1967), 147-171.

2 KOHN, J. J; , "Boundaries of complex manifolds". Proc. Conf. Complex Analysis (Minneapolis 1964) , 81-94. Springer Verlag, Berlin. 1965 .

3 KOHN, J. J. and ROSSI, H., "On the extension of holomorphic functions from the the boundary of a complex manifold". Ann. Math. vol. 81 (1965), 451-472.

4 KOHN, J. J. and NIRENBERG, L., "Non-Coercive boundary value problems". Conn. P. A. Math. vol. 18 (1965), 443-492.

5 KOHN, J. J. and NIRENBERG, L. , "An algebra of pseudo-differential operators". Conn. P. A. vol 18 (1965), 269-305.

CENTRO INTERNAZIONALE MATEMATICO ESTIVO

(C.I.M.E.)

R. SEELEY

TOPICS IN PSEUDO-DIFFERENTIAL OPERATORS

Corso tenuto a Stresa dal 26 Agosto al 3 Settembre 1968

Topics in pseudo-differential operators

R. Seeley*

The subject of pseudo-differential operators has sprung up in the last few years out of the earlier work of Giraud, Mihlin, and Calderon and Zygmund, and is still in the process of development. There are so many contributors to this development that the references given at the article are restricted to those papers actually referred to.

The first four chapters give the elementary theory of pseudo-differential operators, together with some fairly direct applications to elliptic problems on compact manifolds. The last two chapters sketch two more complicated applications, one to the study of the powers of an elliptic operators, and the other to boundary problems.

*Sloan Foundation Fellow; work partially supported by NSF Grant GP6761.

I. Distributions and Sobolef Spaces.

R. Seeley

§1. Tempered distributions and Fourier transforms.

The quickest route into partial differential equations uses Fourier transforms and distributions; we begin with some essential background material on these topics.

Notations:

$$x = (x_1, \dots, x_\nu) \in R^\nu \,, \quad x \cdot y = xy = \langle x,y \rangle = \sum_1^\nu x_j y_j \,, \quad |x| = \langle x,x \rangle^{1/2} \,;$$

$$x^\alpha = x_1^{\alpha_1} \dots x_\nu^{\alpha_\nu}, \text{ where } \alpha = (\alpha_1, \dots, \alpha_\nu) \text{ and each } \alpha_j \text{ is an integer } \geq 0 .$$

For each α we set $\alpha! = \Pi(\alpha_j)!$, and $|\alpha| = \sum_1^\nu \alpha_j$; apparently this does not genera[lly] lead to confusion with the definition of $|x|$.

$$D = -i(\partial/\partial x_1, \dots, \partial/\partial x_\nu) \,, \quad D^\alpha = (-i)^{|\alpha|} \partial^{|\alpha|} / \partial x_1^{\alpha_1} \dots \partial x_\nu^{\alpha_\nu} .$$

$C_c^\infty(\Omega)$ = all complex functions in R^ν with continuous derivatives of all orders, and vanishing outside some compact subset of the open set Ω.

$\mathcal{S}$ = all complex functions φ on R^ν such that for every α and β there is a constant $c_{\alpha\beta\varphi}$ such that $|x^\alpha D^\beta \varphi(x)| \leq c_{\alpha\beta\varphi}$. The convex sets

$$N_{\alpha\beta\varepsilon} = \{\varphi : |x^\alpha D^\beta \varphi(x)| < \varepsilon \text{ for all } x \text{ in } R^\nu\}$$

generate the neighborhoods of zero in $\mathcal{S}$.

It is easy to show that $C_c^\infty(R^\nu)$ is a dense subset of $\mathcal{S}$; if $\varphi \varepsilon \mathcal{S}$, $\psi \varepsilon C_c^\infty(R^\nu)$ and $\psi(0) = 1$, then as $t \longrightarrow 0$, $\psi(tx)\varphi(x) \longrightarrow \varphi(x)$ in $\mathcal{S}$. (More precisely, $\psi(t\cdot) \longrightarrow \varphi(\cdot)$ in $\mathcal{S}$.). The proof of this is very simple, using Leibniz' formula

$$D^\beta(fg) = \sum_{\gamma\leq\beta} C_{\gamma\beta}\, D^{\beta-\gamma}f\, D^\gamma g \;,$$

where $\gamma \leq \beta$ means $\gamma_j \leq \beta_j$ for all j, and $C_{\gamma\beta} = \dfrac{\beta!}{\gamma!(\beta-\gamma)!}$.

Each of the following maps is a continuous transformation on $\mathcal{S}$:

$$\varphi \longrightarrow M_\psi \varphi = \psi\varphi \qquad (\psi \text{ in } \mathcal{S})$$

$$\varphi \longrightarrow M_P \varphi = P\varphi \qquad (P \text{ a polynomial})$$

$$\varphi \longrightarrow D^\alpha \varphi$$

$$\varphi \longrightarrow \psi * \varphi \quad (\text{convolution})$$

where $(\psi * \varphi)(x) = \int \psi(x-y)\varphi(y)dy$, and $\psi \varepsilon \mathcal{S}$. The continuity of this last map can be checked directly, but it also follows easily from properties of the Fourier transform:

$$\hat{\varphi}(\xi) = \int e^{-i\xi\cdot x}\varphi(x)dx \;.$$

For this we have the formulas

$$\widehat{D^\alpha \varphi}(\xi) = \xi^\alpha \hat{\varphi}(\xi)$$

$$\widehat{X^\alpha \varphi} = (-D)^\alpha \hat{\varphi} \qquad (X^\alpha \text{ is the function } X^\alpha(x) = x^\alpha)$$

$$\widehat{\varphi * \psi} = \hat{\varphi}\hat{\psi}$$

$$\widehat{\varphi\psi} = (2\pi)^{-\nu} \hat{\varphi} * \hat{\psi}$$

$$\int \hat{\varphi}\psi = \int \varphi\hat{\psi}$$

$$\varphi(x) = (2\pi)^{-\nu} \int e^{ix\xi} \hat{\varphi}(\xi) d\xi \qquad \text{(Fourier inversion).}$$

$$\int \varphi\bar{\psi} = (2\pi)^{-\nu} \int \hat{\varphi}\bar{\hat{\psi}} \qquad \text{(Parseval's formula)}$$

These formulas are all simple to prove, except the inversion formula, from which Parseval's formula can be derived. (See Hörmander [1].)

The Parseval formula and the inversion formula, coupled with the fact that $\mathcal{S}$ is dense in L^2, lead to the Plancherel theorem: There is a unique continuous extension of the Fourier transform on $\mathcal{S}$ to an isomorphism of L^2 with itself, and with this extension Parseval's formula is valid for φ and ψ in L^2.

The formulas for $\widehat{D^\alpha \varphi}$ and $\widehat{x^\alpha \varphi}$ show that the map $\varphi \longrightarrow \hat{\varphi}$ is continuous on $\mathcal{S}$. We have

$$|\xi^\alpha D^\beta \hat{\varphi}(\xi)| = |\int e^{-ix\xi} [(1+|x|^2)^\nu D^\alpha(-x)^\beta \varphi(x)] (1+|x|^2)^{-\nu} dx|$$

$$\leq [\int (1+|x|^2)^{-\nu} dx] \sup_x |(1+|x|^2)^\nu D^\alpha(-x)^\beta \varphi(x)| ;$$

here $\int (1+|x|^2)^{-\nu}dx < \infty$, and the sup on the right can be made arbitrarily small by restricting φ close to 0 in $\mathcal{S}$.

Because of the inversion formula, the map $\hat{\varphi} \longrightarrow \varphi$ is also continuous, so the Fourier transform is a homeomorphism on $\mathcal{S}$. Corollary: $\varphi \longrightarrow \psi * \varphi$ is continuous on $\mathcal{S}$, since $\widehat{\psi * \varphi} = \hat{\varphi}\hat{\psi}$, and $\hat{\varphi} \longrightarrow \hat{\varphi}\hat{\psi}$ is continuous.

The tempered distributions, denoted $\mathcal{S}'$, are the continuous linear maps of $\mathcal{S} \longrightarrow \mathbb{C}$.

Example 1. If $\int (1+|x|)^{-k} |f(x)|dx < \infty$ for some k, then the map F given by $F(\varphi) = \int \varphi f$ is in $\mathcal{S}'$. When F has such a representation, f is called its density. By standard real variable arguments, f(x) is determined by F for almost all x. For an F of this type we say (loosely) "F is a function".

Example 2. $F(\varphi) = \varphi(0)$. This F is in $\mathcal{S}'$, but it has no density; loosely, F is not a function. Nevertheless, this F is called the "Dirac δ function", denoted δ: $\delta(\varphi) = \varphi(0)$. More generally, $\delta_x(\varphi) = \varphi(x)$.

Operations on $\mathcal{S}'$ are motivated by the case where F has a density f in $\mathcal{S}$; the operation on F should reduce to the same operation performed on f. For instance, the density of $D^\alpha F$ should be $D^\alpha f$. Since $\int \varphi D^\alpha f = (-1)^{|\alpha|}\int f D^\alpha \varphi = (-1)^{|\alpha|}F(D^\alpha \varphi)$, we make the following definition of differentiation for a general F in $\mathcal{S}'$:

$$D^\alpha F(\varphi) = (-1)^{|\alpha|}F(D^\alpha \varphi).$$

Similarly, we define

$$\hat{F}(\varphi) = F(\hat{\varphi})$$

$$aF(\varphi) = F(a\varphi) \quad \text{if } a \in \mathcal{S}\text{, or } a \text{ is a polynomial.}$$

Clearly, the Fourier transform is a homeomorphism on $\mathcal{S}'$, and multiplication by a is continuous on $\mathcal{S}'$. Convolution can be defined like this: For φ in $\mathcal{S}$, set $\tau_x\varphi(y) = \varphi(x-y)$; then for F in $\mathcal{S}'$, we have the C^∞ function:

$$F*\varphi(x) = \varphi*F(x) = F(\tau_x\varphi).$$

The formulas for Fourier transforms in $\mathcal{S}$ extend directly to $\mathcal{S}'$ (when they make sense). For instance,

$$\widehat{\varphi F} = (2\pi)^{-\nu}\hat{\varphi}*F\text{ , } \varphi \text{ in } \mathcal{S}\text{, } F \text{ in } \mathcal{S}'\text{ .}$$

Example 3. $\hat{\delta}(\varphi) = \delta(\hat{\varphi}) = \hat{\varphi}(0) = \int\varphi$; $\hat{\delta}_x(\varphi) = \hat{\varphi}(x) = \int e^{-x\xi i}\varphi(\xi)d\xi$

$\delta*\varphi(x) = \varphi(x)$, $\delta_y*\varphi(x) = \varphi(x-y)$.

§2. The Sobolef Spaces $H^s(R^\nu)$.

The distributions we need primarily are those in the Sobolef spaces H^s.

Definition. Let s be real. $H^s(R^\nu) = \{F \text{ in } \mathcal{S}' : \hat{F}$ has a density g such that $\int(1+|\xi|^2)^s \, |g(\xi)|^2 d\xi < \infty\}$. For such F, the s-norm is

$$||F||_s = (2\pi)^{-\nu/2} \left(\int(1+|\xi|^2)^s \, |g(\xi)|^2 d\xi\right)^{1/2} .$$

Example 4. From Example 3, $\hat{\delta}$ has the density $\hat{g}(\xi) = 1$, so $\delta \in H^s$ for all $s < -\nu/2$. More generally, $\hat{\delta}_x$ has the density $g(\xi) = e^{-ix\xi}$, and it follows from the dominated convergence theorem that the map $x \longrightarrow \delta_x$ is continuous: $R^\nu \longrightarrow H^s$, $s < -\nu/2$.

For convenience, we denote the density of the distribution $\hat{F}$ by the same symbol $\hat{F}$. Thus $(||F||_s)^2 = (2\pi)^{-\nu} \int(1+|\xi|^2)^s \, |\hat{F}(\xi)|^2 d\xi$.

From the definition we have $H^s \supset H^t$ if $s < t$, and the inclusion has norm 1.

When $F \in H^s$, $s \geq 0$, then $\hat{F} \in L^2$, so by the Plancherel theorem there is an f in L^2 with $\hat{f} = \hat{F}$. It is easy to check that f is a density of F, so all our formulas are consistent. In particular, H^0 is isometric with L^2, under the correspondence $F \longleftrightarrow f$ = density of F.

It is clear that H^s is a Hilbert space, but we seldom use this fact for $s \neq 0$. However, we do use the pairing for F in H^s, G in H^{-s}, given by

$$< F,G > = (2\pi)^{-2} \int \hat{F}(\xi) \hat{G}(-\xi) d\xi$$

which provides a natural identification of H^{-s} with the dual of H^s.

From the formula

$$\widehat{D^\alpha F} = \Xi^\alpha \hat{F} \qquad (\text{where } \Xi^\alpha(\xi) = \xi^\alpha),$$

it follows easily that $D^\alpha : H^s \longrightarrow H^{s-|\alpha|}$ has norm 1. In particular, if $F \varepsilon H^k$ for some integer $k \geq 0$, then $D^\alpha F \varepsilon H^0$ for $|\alpha| \leq k$, so (by Plancherel) $D^\alpha F$ has a density f_α in L^2. In fact, this property characterizes H^k for integer $k \geq 0$. Precisely, we have for certain constants $C_\alpha \geq 0$

$$(1+|\xi|^2)^k = \sum_{|\alpha| \leq k} C_\alpha \xi^{2\alpha},$$

so for F in H^k

$$(2\pi)^n \, ||F||_k^2 = \int (1+|\xi|^2)^k \, |\hat{F}(\xi)|^2 d\xi$$

$$= \Sigma \, C_\alpha \int \xi^{2\alpha} \, |\hat{F}(\xi)|^2 d\xi = \Sigma \, C_\alpha ||D^\alpha F||_0^2 = \Sigma C_\alpha ||f_\alpha||_{L^2}^2 .$$

Similarly, we can characterize H^{-k} as all sums of the form $\sum_{|\alpha| \leq k} D^\alpha F_\alpha$, where $F_\alpha \varepsilon H^0$; given F in H^{-k}, we define F_α by $\widehat{F_\alpha}(\xi) = C_\alpha \, \xi^\alpha (1+|\xi|^2)^{-k} \hat{F}(\xi)$, and it follows easily that $F_\alpha \varepsilon H^0$, $F = \Sigma \, D^\alpha F_\alpha$, and $||F||_{-k}^2 = \Sigma \, ||F_\alpha||_0^2$.

$\mathcal{S}$ is dense in H^s for every s, in the following sense:

Theorem 1. To each f in $\mathcal{S}$ associate the distribution F in $\mathcal{S}'$ defined by $F(\varphi) = \int f\varphi$. Then for every s, F lies in H^s, and the map $f \longrightarrow F$ injects $\mathcal{S}$ continuously onto a dense subset of H^s.

Proof. It is clear that $f \longrightarrow F$ is a continuous injection, since $\hat{F}(\xi) = \hat{f}(\xi) \in \mathcal{S}$. To prove the density, suppose $G \in H^s$ and $\varepsilon > 0$ are given, and set $h(\xi) = (1+|\xi|^2)^{s/2}\hat{G}(\xi)$. Then h is in L^2, so there is an h_ε in $\mathcal{S}$ with $||h_\varepsilon - h||_{L^2} < (2\pi)^\nu \varepsilon$. If g_ε is the inverse Fourier transform of $(1+|\xi|^2)^{-s/2} h_\varepsilon(\xi)$ and G_ε the corresponding distribution, then $||G_\varepsilon - G||_s < \varepsilon$.

Corollary 1. $C_c^\infty(R^\nu)$ is dense in H^s, since it is dense in $\mathcal{S}$.

Exemple 5. Let $f \in C_c^\infty$, $\int f = 1$, and set $f_n(x) = n^\nu f(nx)$. Then if F_n is the corresponding distribution, $F_n \longrightarrow \delta$ in H^s for every $s < -\nu/2$. This follows since

$$\hat{f}_n(\xi) = \hat{f}(\xi/n) \longrightarrow \hat{f}(0) = \int f = 1 ,$$

$$\hat{f}_n \text{ is bounded, } \hat{\delta}(\xi) = 1 , \quad \int (1+|\xi|^2)^s d\xi < \infty ,$$

so

$$(2\pi)^\nu ||F_n - \delta||_s^2 = \int (1+|\xi|^2)^s \, |\hat{f}(\xi/n) - 1|^2 d\xi \longrightarrow 0,$$

by the dominated convergence theorems.

From Theorem 1 follows the

<u>Corollary 2.</u> If $f \in \mathcal{S}$, $G \in H^s$, and $F(\varphi) = \int f\varphi$, then $F \in H^{-s}$ and $\langle G,F \rangle = G(f)$.

<u>Proof.</u> By density, we need only consider G of the form $G(\varphi) = \int g\varphi$ for a g in $\mathcal{S}$. In this case $\hat{G}(\xi) = \hat{g}(\xi)$, and

$$< G,F > = (2\pi)^{-\nu} \int \hat{g}(\xi)\hat{f}(-\xi)d\xi = \int gf = G(f).$$

From this Corollary we obtain the most elementary version of Sobolef's theorem:

<u>Theorem 2.</u> If $F \in H^s$, $s > \nu/2$, then F has the continuous bounded density $f(x) = < F, \delta_x >$.

<u>Proof.</u> For φ in $\mathcal{S}$, the map $x \longrightarrow \varphi(x)\delta_x$ is continuous from R^ν into H^{-s}, and has an absolutely convergent improper Riemann integral, which is evaluated as follows:

$$(\int\varphi(x)\delta_x dx)(\psi) = \int\varphi(x)\delta_x(\psi)dx = \int\varphi(x)\psi(x)dx ,$$

i.e. if Φ is the distribution with density φ, then

$$\int\varphi(x)\delta_x dx = \Phi . \qquad (*).$$

Hence by Corollary 2,

$$F(\varphi) = \langle F, \bar{\varphi} \rangle = \langle F, \int \varphi(x)\delta_x dx \rangle = \int \langle F, \delta_x \rangle \varphi(x) dx.$$

<u>Corollary.</u> If $F \varepsilon H^{s+k}$, $s > \nu/2$, then F has a density of class C^k.

<u>Proof.</u> For $|\alpha| \leq k$ we have $D^\alpha F \varepsilon H^s$, so $f_\alpha(x) = \langle D^\alpha F, \delta_x \rangle$ is continuous. Finally, $f_\alpha = D^\alpha f$, since

$$\int f_\alpha \varphi = D^\alpha F(\varphi) = (-1)^{|\alpha|} F(D^\alpha \varphi) = (-1)^{|\alpha|} \int f D^\alpha \varphi .$$

<u>Remark:</u> It is not hard to show that f is <u>Holder continuous</u>, precisely that $f \varepsilon C^{s-\nu/2}$ if $s-\nu/2$ is positive and not an integer. This can be proved by taking Fourier transforms and applying the Schwartz inequality.

A result similar to Sobolef's theorem is

<u>Theorem 3.</u> If A is a bounded linear operator from H^{-s} to H^s for some $s > \nu/2$, then A has the continuous bounded kernel $K(x,y) = \langle A\delta_y, \delta_x \rangle$. By this we mean that if $F \varepsilon H^s$ has a density f in $\mathcal{S}$, then Af has the density $\int K(x,y) f(y)$

<u>Proof.</u> Since $y \longrightarrow \delta_y$ is continuous and bounded into H^{-s}, the function $\langle A\delta_y, \delta_x \rangle$ is bounded and continuous. Further, for $f \varepsilon \mathcal{S}$, $\varphi \varepsilon \mathcal{S}$, we have

$$\iint < A\delta_y, \delta_x > f(y)dy\, \omega(x)dx = < A\int f(y)\delta_y dy, \int \delta_x \omega(x)dx >$$

$$= < AF, \phi > = AF(\omega) ,$$

by (*) and Corollary 2 of Theorem 1 above.

Corollary. If A is bounded from H^{-s-k} to $H^{s+\ell}$ for $s > \nu/2$, then the kernel $K(x,y)$ has bounded continuous derivatives $(D_x)^\alpha (D_y)^\beta K(x,y)$ for $|\alpha| \leq \ell$, $|\beta| \leq k$.

Proof. These derivatives are simply the kernels of $(-1)^{|\beta|} D^\alpha A\, D^\beta$.

Remark. It should be clear that Theorem 2 does not characterize H^s, and Theorem 3 does not characterize the bounded operators from H^{-s} to H^s. However, we can give a partial converse of Theorem 3.

Theorem 4. If $K_{\alpha\beta}(x,y) = D_x^\alpha D_y^\beta K(x,y)$ is continuous for $|\alpha| \leq \ell$, $|\beta| \leq k$, and there is a constant C such that

$$\int |K_{\alpha\beta}(x,y)|dx \leq C , \quad \int |K_{\alpha\beta}(x,y)|dy \leq C$$

for these α and β, then

$$Af(x) = \int K(x,y)f(y)dy$$

defines a bounded operator from H^{-k} to H^{ℓ}.

The proof of this uses:

<u>Lemma 1.</u> If $\int |K(x,y)|dy \leq C$ and $\int |K(x,y)|dx \leq C$, then K defines an operator on L^2 with norm $\leq C$.

<u>Proof.</u> $$\left|\iint K(x,y)f(y)dy\; g(x)dx\right|$$

$$\leq \iint |K(x,y)|^{1/2}|f(y)|\;|K(x,y)|^{1/2}\;|g(x)|dy\,dx$$

$$\leq \left(\iint |K(x,y)|\;|f(y)|^2 dx\,dy\right)^{1/2}\left(\iint |K(x,y)|\;|g(x)^1 dy\,dx\right)^{1/2}$$

$$\leq C\|f\|_{L^2}\;\|g\|_{L^2}\,.$$

To prove Theorem 4, let $f, g \in \mathcal{S}$, and let F,G be the corresponding distributions. Then, as we have seen in characterizing H^{-k}, there are functions f_α, g_β, such that $f = \sum_{|\alpha|\leq k} D^\alpha f_\alpha$, $g = \sum_{|\beta|\leq \ell} D^\beta g_\beta$, and .

$$\|\;\|^2_{-k} = \Sigma\|f_\alpha\|^2_{L^2}\,,\quad \|G\|^2_{-\ell} = \Sigma\|g_\beta\|^2_{L^2}\,.$$

It is easy to check, from their construction, that $f_\alpha, g_\beta \in \mathcal{S}$, so we find

$$< AF,G > = \iint K(x,y)f(y)g(x)dy\,dx$$

$$= \Sigma\Sigma\iint K(x,y)D^\alpha f_\alpha(y)\ D^\beta g_\beta(x)dy\,dx$$

$$= \Sigma\Sigma(-1)^{|\alpha|+|\beta|}\iint K_{\alpha\beta}(x,y)f_\alpha(y)g_\beta(x)dy\,dx .$$

Because of the conditions on $K_{\alpha\beta}$, we obtain as before

$$|<AF,G>| \leq C\ \Sigma\Sigma||f_\alpha||\cdot||g_\beta|| \leq C'||F||_{-k}\ ||G||_{-\ell} .$$

Because of the duality between H^s and H^{-s}, we get $||AF||_\ell \leq C'||F||_{-k}$ for F with a density f in $\mathcal{S}$; then, by passage to the limit, we get $||AF||_\ell \leq C'||F||_{-k}$ for all F in H^{-k}, and Theorem 4 is proved.

Interpolation theorems like the following one are a powerful tool in working with Sobolef spaces.

Theorem 5. Suppose A is linear from $\mathcal{S}(R^\nu)$ to $\mathcal{S}'(R^\mu)$, and

$$||A\varphi||_{t_j} \leq C_j ||\varphi||_{s_j}, \quad j = 0,1.$$

Let $s(\theta) = (1-\theta)s_0 + \theta s_1$, $t(\theta) = (1-\theta)t_0 + \theta t_1$, $0 \leq \theta \leq 1$. Then

$$||A\varphi||_{t(\theta)} \leq C_0^{1-\theta} C_1^{\theta} ||\varphi||_{s(\theta)} .$$

Proof. We use the classical Phragmen-Lindelof theorem: If $F(z)$ is bounded and analytic for $0 \leq \mathrm{Re}(z) \leq 1$, and $|F(iy)| \leq C_0$, $|F(1+iy)| \leq C_1$, then $|F(x+iy)| \leq C_0^{1-x} C_1^{x}$. This is a slight extension of the maximum principle; for a proof and other applications, see Calderon and Zygmund [1]. For our application we introduce an operator Λ^ω for ω complex by

$$\widehat{\Lambda^\omega f}(\xi) = (1+|\xi|^2)^{\omega/2} \hat{f}(\xi).$$

Then Λ^{ω} is an isometry of H^s onto $H^{s-\mathrm{Re}(\omega)}$. Now let f be in $\mathcal{S}(R^{\nu})$, g in $\mathcal{S}(R^{\mu})$, and consider

$$F(z) = < A\Lambda^{-s(z)}f,\ \Lambda^{t(z)}g > .$$

By the hypotheses on A, F is bounded and analytic for $0 \leq \mathrm{Re}(z) \leq 1$, and

$$|F(iy)| \leq C_0 ||\Lambda^{-s(iy)}f||_{s_0}\ ||\Lambda^{t(iy)}g||_{-t_0}$$

$$= C_0 ||\Lambda^{-s_0}f||_{s_0} ||\Lambda^{t_0}g||_{-t_0} = C_0 ||f||_0 ||g||_0 .$$

Similarly, $|F(1+iy)| \leq C_1 ||f||_0\ ||g||_0$, so for $0 \leq \theta \leq 1$ we have

$$|\langle A\Lambda^{-s(\theta)}f,\ \Lambda^{t(\theta)}g\rangle| \leq C_0^{1-\theta}\ C_1^{\theta}\ ||f||_0\ ||g||_0$$

$$= C_0^{1-\theta}C_1^{\theta}\ ||\Lambda^{-s(\theta)}f||_{s(\theta)}\ ||\Lambda^{t(\theta)}g||_{-t(\theta)} .$$

Since $\Lambda^{-s(\theta)}f$ and $\Lambda^{t(\theta)}g$ run through $\mathcal{S}$ as f and g do, we get

$$||A\varphi||_{t(\theta)} \leq C_0^{1-\theta} C_1^{\theta} \, ||\varphi||_{s(\theta)} \, .$$

Corollary 1. If φ and all its derivatives of order $\leq k$ are bounded on R^ν, then $F \longrightarrow \varphi F$ is continuous from H^s to H^s for $|s| \leq k$.

Proof. The result holds for $s = k$ by the characterization of H^k for $k =$ integer ≥ 0, then for $s = -k$ by duality, and finally for $-k \leq s \leq k$ by interpolation, i.e. by Theorem 5.

Corollary 2. If $\chi: U \longrightarrow V$ is a diffeomorphism of open sets in R^ν, of class C^k with $k \geq 1$, and if $\varphi \in C_c^\infty(U)$, then the map A,

$$Af(x) = \begin{cases} \varphi(x) f(\chi(x)), & x \in U \\ 0 & x \notin U \end{cases}$$

is bounded on H^s for $1 - k \leq s \leq k$.

The proof is just like Corollary 1. Notice that the Jacobian determinant is class C^{k-1}; this explains the range $1-k \leq s \leq k$.

Theorem 6. (Rellich). Let $\varphi \varepsilon \mathscr{S}$, and define an operator M_φ on $\mathscr{S}$ by $M_\varphi F = \varphi F$. Then M_φ is compact from H^s to H^t for $t<s$.

Proof. Suppose a sequence F_n is given with $||F_n||_s = 1$. Set $g_n(\xi) = \widehat{M_\varphi F_n}(\xi) = \varphi * \hat{F}_n(\xi)$. Then for ξ in any compact set K,

$$|g_n(\xi)|^2 \leq \int (1+|\eta|^2)^{-s} |\hat{\varphi}(\xi-\eta)|^2 d\eta \int (1+|\eta|^2)^s |\hat{F}_n(\eta)|^2 d\eta$$

$$\leq C_K ,$$

and similarly

$$|D^\alpha g_n(\xi)|^2 = |D^\alpha \hat{\varphi} * \hat{F}_n(\xi)|^2 \leq C_K ,$$

so g_n is a uniformly bounded, equicontinuous sequence on every compact K, and by passing to a subsequence we may assume that g_n converges to a function g, uniformly on every compact subset. By Corollary 1 of Theorem 5,

$$\int |g_n(\xi)|^2(1+|\xi|^2)^s d\xi = ||M_\varphi F_n||_s^2 \leq C||F_n||_s^2 = C,$$

so

$$\int |g(\xi)|^2(1+|\xi|^2)^s d\xi \leq C.$$

Now set $G(\hat{\psi}) = \int g\psi$. Then $\hat{G} = g$, so .

$$||\varphi F_n - G||_t^2 = \int (1+|\xi|^2)^{t-s}|g_n(\xi) - g(\xi)|^2(1+|\xi|^2)^s d\xi. \qquad (**)$$

Split this integral into two parts, one for $|\xi| \geq R$ and one for $|\xi| \leq R$. Since $\int |g_n(\xi) - g(\xi)|^2(1+|\xi|^2)^s d\xi \leq$ const and $(1+|\xi|^2)^{t-s} \longrightarrow 0$ as $|\xi| \longrightarrow \infty$, we can for any $\varepsilon > 0$ choose R so that the integral (**) for $|\xi| \geq R$ is $< \varepsilon/2$; then, since $g_n \longrightarrow g$, we can choose n so large that the integral for $|\xi| \leq R$ is $< \varepsilon/2$. It follows that $||\varphi F_n - G||_t \longrightarrow 0$, and the theorem is proved.

II. Pseudo-differential operators on R^{ν}

There are many different definitions of pseudo-differential operators (ψdo's), all depending on the inclinations of the expositor, and on what results he has in mind. The course taken here is a blend of Kohn-Nirenberg [1] and Hörmander [3].

A differential operator $Af(x) = \sum_{|\alpha|\leq k} a_{\alpha}(x)D^{\alpha}f(x)$ can be written

$$Af(x) = (2\pi)^{-\nu}\int e^{ix\xi}\, a(x,\xi)\hat{f}(\xi)d\xi,$$

where $a(x,\xi) = \sum_{|\alpha|\leq k} a_{\alpha}(x)\xi^{\alpha}$ is the symbol, or (complete) characteristic polynomial of A, denoted $\sigma(A)$.

To obtain ψdo's we replace the polynomial a by a function of more general type.

Definition 1. Let m be real. S^m consists of all functions $a(x,\xi)$ such that for every α,β,γ, $\sup_{x,\xi}(1+|\xi|)^{|\gamma|-m}|x^{\alpha}D^{\beta}D_{\xi}^{\gamma}\, a(x,\xi)| < \infty$. These seminorms make S^m a Frechet space.

This is almost the same as the class $S^m_{1,0}$ given in Hormander's article just cited, except that we have added some control over the behavior as $x \longrightarrow \infty$, as Kohn-Nirenberg do.

Definition 2. For a in S^m, we define Op(a) (also denoted a(x,D)) by

$$Op(a)f(x) = (2\pi)^{-\nu} \int e^{ix\xi} \hat{f}(\xi)\, a(x,\xi) d\xi\,, \quad f \text{ in } \mathcal{S}.$$

The class of all such operators is denoted $Op(S^m)$.

Notice that if A is a differential operator with $\sigma(A) = a$, we have $A = Op(a)$.

Definition 3. If a is in some S^k and $a_{m-j} \varepsilon S^{m-j}$, $j = 0,1,2,\ldots$, we write $a \sim \Sigma a_{m-j}$ if and only if $a - \sum_{j<r} a_{m-j}$ is in S^{m-r}. In particular, $a \sim b$ if a-b is in $\bigcap_{j=1}^{\infty} S^{-j}$.

Definition 4. Let ω be complex, and $m = Re(\omega)$. SH^{ω} is the subspace of S^m consisting of functions a which have a homogeneous asymptotic expansion

$a \sim \Sigma a_{\omega-j}$, where

$$a_{\omega-j}(x, t\xi) = t^{\omega-j} a(x,\xi) \text{ for } t \geq 1, \ |\xi| \geq c .$$

Definition 5. Let A be a linear operator from $\mathcal{S}(R^\nu)$ to $\mathcal{S}'(R^\nu)$. Then A has order $\leq m$ iff, for real s, $\|A\|_{s,s-m} = \sup_{f \neq 0} \frac{\|Af\|_{s-m}}{\|f\|_s} < \infty$. The collection of these norms defines the topology of operators of order $\leq m$.

Thus, for example, $f \longrightarrow \varphi f$ has order $\leq$ zero if φ and all its derivatives are bounded (Theorem I.5, Corollary 1); $f \longrightarrow D^\alpha f$ has order $\leq |\alpha|$; and $f \longrightarrow \Lambda^\omega f$ has order $\mathrm{Re}(\omega)$, as we noted in proving Theorem I.5.

It is easy to prove that Op(a) is continuous from $\mathcal{S}$ into $\mathcal{S}$. Further:

Theorem 1. For a in S^m, Op(a) has order $\leq m$, and the map $a \longrightarrow \mathrm{Op}(a)$ is continuous.

Proof. Let $\hat{a}(\eta, \xi) = \int e^{-ix\eta} a(x,\xi)dx$. Then

$$\widehat{Op(a)f}(\eta) = (2\pi)^{-\nu} \int \hat{a}(\eta-\xi,\xi)\,\hat{f}(\xi)d\xi . \qquad (1)$$

From the definition of the norm in H^s, we have to estimate the norm of the operator $\Lambda^{s-m}Op(a)\Lambda^{-s}$ on L^2. Taking Fourier transforms of this, we obtain the operator with kernel

$$b(\eta,\xi) = k(\eta)^{s-m}\hat{a}(\eta-\xi,\xi)\,k(\xi)^{-s} ,$$

where

$$k(\eta) = (1+|\eta|^2)^{1/2} . \qquad (2)$$

From the conditions on a (Definition 1) we have for each N a constant C_N such that

$$|\hat{a}(\eta-\xi,\xi)| \leq C_N\,k(\eta-\xi)^{-N}k(\xi)^m . \qquad (3)$$

From the "Peetre's" inequality

$$k(\xi)^t \leq 2^{|t|}k(\eta)^t\,k(\xi-\eta)^{|t|} \qquad (4)$$

we get

$$|b(\tau,\xi)| \leq C'_N\, k(\tau-\xi)^{|s-m|-N} .$$

With N sufficiently large, we get bounds on $\int|b(\eta,\xi)|d\eta$ and $\int|b(\tau,\xi)|d\xi$, hence the Theorem follows from Lemma I.1 and the Plancherel theorem.

Theorem 2. For a in S^m and b in S^n, we have

$$Op(b)\ Op(a) = Op(c) ,$$

where c is in S^{m+n} and

$$c \sim \sum_k \sum_{|\alpha|=k} (iD_\xi^\alpha b)\,(D_x)^\alpha a/\alpha! .$$

Proof. Using formula (1), we get Op(b) Op(a) = Op(c), where

$$c(x,\xi) = (2\pi)^{-\nu} \int e^{ix(\eta-\xi)} b(x,\eta)\,\hat{a}(\eta-\xi,\xi)d\eta .$$

$$= (2\pi)^{-\nu} \int e^{ix\sigma}\, b(x,\xi+\sigma)\,\hat{a}(\sigma,\xi)d\sigma . \qquad (5)$$

Using (3), the last integrand is dominated by

$$k(\xi+\sigma)^{n}\,[k(\xi)^{m}k(\sigma)^{-N}] \leq 2^{|n|}k(\xi)^{m+n}k(\sigma)^{|n|-N}\ ,$$

which with $N > |n|+\nu$ shows that

$$|c(x,\xi)| \leq C(1+|\xi|)^{m+n}\ .$$

Similar estimates provide bounds for $x^{\alpha}D_x^{\beta}D_{\xi}^{\nu}c$, hence c is in S^{m+n}.

To obtain the asymptotic form of c, we use a Taylor expansion of $b(x,\xi+\sigma)$ in (5),

$$b(x,\xi+\sigma) = \sum_{|\alpha|<r} b^{(\alpha)}(x,\xi)\sigma^{\alpha}/\alpha! \ + R\ ,$$

$$R(x,\xi,\sigma) = r\int_0^1 (1-t)^{r-1} \sum_{|\alpha|=r} b^{(\alpha)}(x,\xi+t\sigma)\sigma^{\alpha}/\alpha!\ dt\ ,$$

$$b^{(\alpha)}(x,\xi) = (\partial/\partial\xi)^{\alpha}b(x,\xi)\ .$$

The estimate

$$|x^\beta D_x^\gamma D_\xi^\delta b^{(\alpha)}(x,\xi+t\sigma)| \leq C\,k(\xi+t\sigma)^{m-|\alpha|-|\delta|}$$

$$\leq C\,k(\xi)^{m-|\alpha|-|\delta|}\,k(\sigma)^{|(m-|\alpha|-|\delta|)|}$$

shows that $c(x,\xi) - \sum_{|\alpha|<r} b^{(\alpha)} D_x^\alpha a/\alpha!$ is in S^{m+n-r}, and Theorem 2 is proved.

__Definition 6.__ The function c in formula (5) is denoted $b \circ a$.

__Corollary.__ From the proof of Theorem 2, the map $(b,a) \longrightarrow b\circ a - \sum_{|\alpha|<r} b^{(\alpha)} D_x^\alpha a/\alpha!$ is continuous from $S^n \times S^m$ to S^{m+n-r}, for each $r = 0,1,2,\ldots$.

__Theorem 3.__ For __a__ in S^m, there is an a' in S^m such that

$$\int [Op(a)f]\,g = \int f\,Op(a')g\,, \qquad f,g \text{ in } S, \tag{6}$$

and the map $a \longrightarrow a'(x,\xi) - \sum_{|\alpha|<r} (-D_x^\alpha)\, a^{(\alpha)}(x,-\xi)$ is continuous from S^m to S^{m-r} for $r = 0,1,2,\ldots$.

Proof. Set

$$a'(y,\eta) = (2\pi)^{-\nu}\int e^{iy(\xi-\eta)}\hat{a}(\xi-\eta,-\xi)d\xi . \qquad (7)$$

It is easy to check that this a' is in S^m and has the desired asymptotic expansion, much as in Theorem 2, by expanding $\hat{a}(\sigma,-\xi)\sim\sum\hat{a}^{(\alpha)}(\sigma,-\eta)(\eta-\xi)^\alpha/\alpha!$ It remains only to show that $Op(a')$ is the dual of $Op(a)$. Suppose at first that $a(x,\xi)$ vanishes for large ξ. Then from (7) and Fubini's theorem,

$$\int f\,Op(a')g = \iiint e^{i(y-x)\eta}\,a'(y,\eta)\,g(x)dx\,d\eta\,f(y)dy$$

$$= \iiint [(2\pi)^{-\nu}\int e^{ix(\xi-\eta)}\hat{a}(\xi-\eta,-\xi)d\eta]\,e^{i(y-x)\xi}\,fg\,dy\,d\xi\,dx$$

$$= \iiint e^{i(y-x)\xi}\,\hat{a}(x,-\xi)f(y)dy\,d\xi\,g(x)dx = \int[Op(a)f]g ,$$

as desired. In general, for a in S^m we take θ in C_c^∞ such that $\theta=1$ in a neighborhood of zero, and set $a_n(x,\xi) = a(x,\xi)\,\theta(\xi/n)$. Then $a_n \longrightarrow a$ and $a_n(x,\xi)$ is dominated by $(1+|\xi|)^m$, so $\int[Op(a_n)f]g \longrightarrow \int[Op(a)f]g$ for f,g in $\mathcal{S}$. Similarly, $\int f\,Op(a'_n)g \longrightarrow \int f\,Op(a')g$, so we obtain (6).

Remark 1. Similarly, $\int [Op(a)f]\bar{g} = \int f\, \overline{Op(a^*)g}$ with a^* in S^m,

$a^* \sim \sum D_x^\alpha (iD_\xi)^\alpha\, \bar{a}/\alpha!$

Remark 2. By Theorem 3, $Op(a)$ can be extended as a map of $\mathcal{S}'$ into $\mathcal{S}'$ by setting $[Op(a)F](\varphi) = F(Op(a')\varphi)$ for all F in $\mathcal{S}'$, φ in $\mathcal{S}$.

Remark 3. It might seem simpler to define S^m (Definition 1) by requiring that $a(x,\xi)$ have compact support in x, rather than rapid decay at ∞; but if we had done that, the statement of Theorem 3 would have been a little more complicated. And if we had made no requirement at ∞, Theorem 2 would be more complicated.

Now we generalize by considering operators defined only locally, and restrict by considering only kernels with a homogeneous expansion. The point of localizing is that pseudo-differential operators serve mainly to study local problems. We take kernels with a homogeneous expansion because this is the class that arises naturally in the problems we will consider here, and because this class can be studied in more detail.

Definition 7. Let A be a linear map: $C_c^\infty(U) \longrightarrow C^\infty(U)$. A is a pseudo-differential operator of degree ω (ψdo_ω) iff given φ and ψ in $C_c^\infty(U)$, $\varphi A \psi$ is in $Op(SH^{\omega})$.

Recalling Definition 4, A is a ψdo_ω iff $\varphi A \psi = Op(a)$ with $a \sim \Sigma a_{\omega - j}$. This is the class considered by Hörmander [2], except that his degrees of homogeneity are real, and are not all equal mod 1.

If we had required $\varphi A \psi$ to be in $Op(S^m)$, we would have the operators of type (1,0) considered in Hormander [3].

<u>Definition 5.</u> If A is a ψdo_ω; if $\varphi, \psi \in C_c^\infty(U)$ and both equal 1 in an open $V \subset U$; and if $\varphi A \psi = Op(a)$, then for x in V we set

$$\sigma(A)(x,\xi) = \Sigma\, a_{\omega - j}(x,\xi), \qquad \xi \neq 0$$

where the "sum" on the right is the asymptotic expansion of a, altered for small ξ so as to be homogeneous for $\xi \neq 0$.

The definition is justified by

<u>Lemma 1.</u> If $a \sim \Sigma a_{\omega - j}$ and $u \in C_c^\infty(R^\nu)$, then

$$e^{-i\langle x,\xi\rangle}[Op(a)\, u\, e^{i\langle\cdot,\xi\rangle}](x) \sim \sum_{\alpha, j} a^{(\alpha)}_{\omega - j}(x,\xi) D^\alpha u(x) \,, \tag{1}$$

where $a^{(\alpha)} = (\partial/\partial\xi)^\alpha a$.

Proof. The left-hand side of (1) equals

$$(2\pi)^{-\nu} \int e^{ix\sigma} a(x,\xi+\sigma)\, \hat{u}(\sigma) d\sigma .$$

Using a Taylor expansion of $a(x,\xi+\sigma)$ in powers of σ, and estimating the remainder as in the proof of Theorem 2 above, we get

$$e^{-i\langle x,\xi\rangle}[Op(a)ue^{i\langle\cdot,\xi\rangle}](x) \sim \sum_{\alpha} a^{(\alpha)}(x,\xi) D^{\alpha} u(x) ,$$

where the asymptotic expansion is in the sense of Definition 3. Now the Lemma follows from $a^{(\alpha)}(x,\xi) \sim \Sigma\, a^{(\alpha)}_{\omega-j}(x,\xi)$.

Now to justify Definition 8, we take φ, ψ, A and $\underline{a}$ as given there, and take u in $C_c^{\infty}(V)$ such that $u \equiv 1$ in a neighborhood of some point x_0 in V. Then for x near x_0 we have $D^{\alpha}u = 0$ for $\alpha \neq 0$, and $\varphi A \psi u e^{i\langle\cdot,\xi\rangle} = Aue^{i\langle\cdot,\xi\rangle}$. Thus for x near x_0,

$$\Sigma\, a_{\omega-j}(x,\xi) \sim e^{-ix\xi}[Op(a)ue^{i\langle\cdot,\xi\rangle}](x) = e^{-ix\xi}[Aue^{i\langle\cdot,\xi\rangle})](x) ;$$

hence the asymptotic expansion depends only on A, not on φ or ψ.

For formal sums $a = \sum a_{\omega-j}$, $b = \sum b_{\mu-k}$, we introduce

$a \circ b = \sum (D_\xi^\alpha a)(iD_x)^\alpha b/\alpha!$, where it is understood that the terms

$$(D_\xi^\alpha a_{\omega-j})(iD_x)^\alpha b_{\mu-k}/\alpha!$$

are to be collected according to the degree of homogeneity in ξ. Since there are only finitely many terms of a given degree of homogeneity, the operation makes sense. Notice that the top term in $a \circ b$ is $a_\omega b_\mu$.

Similarly, we define, for $a = \sum a_{\omega-j}$,

$$a'(x,\xi) = \sum (-iD_x)^\alpha (D_\xi^\alpha a)(x,-\xi)/\alpha!$$

$$a^*(x,\xi) = \sum (iD_x)^\alpha D_\xi^\alpha \bar{a}(x,\xi)/\alpha!$$

For locally defined ψdo's we have

Theorem 4. If A is a ψdo, then there are unique ψdo's A' and A^* such that $\int (Af)g = \int fA'g$, $\int (Af)\bar{g} = \int f\overline{A^* g}$, f, g in $C_c^\infty(U)$; and

$\sigma(A') = \sigma(A)'$, $\sigma(A^*) = \sigma(A)^*$.

If B is a ψdo and $\varphi \in C_c^\infty(U)$, then $B\varphi A$ is a ψdo and

$$\sigma(B\varphi A) = \sigma(B) \circ \sigma(\varphi A) .$$

In particular, on any open set where $\varphi = 1$,

$$\sigma(\varphi) = \sigma(B) \circ \sigma(A).$$

Given any sequence $a_\omega, a_{\omega-1}, \ldots$, where $a_{\omega-j}(x,\xi)$ is C^∞ in $U \times R^\nu$, and homogeneous of degree $\omega-j$ in ξ for large ξ, there is a ψdo A such that $\sigma(A) \sim \Sigma\, a_{\omega-j}$.

Proof. All but the last statement follow from Theorems 2 and 3. For the last statement we use

Lemma 2. Given a_ω, $a_{\omega-1}, \ldots$, as above, there is an $a(x,\xi)$ such that for each J, α, β, and each compact subset K of U,

$$(1+|\xi|)^{-\mathrm{Re}(\omega)-J-1+|\beta|} D_x^\alpha D_\xi^\beta \left[a(x,\xi) - \sum_0^J a_{\omega-j}(x,\xi)\right] \text{ is bounded on } K \times R^\nu .$$

To prove the Lemma, choose a monotone sequence of compact sets K_m such that $\cup K_m = U$, choose a C^∞ function $\theta(\xi)$ which vanishes for small ξ and equals 1 for large ξ, and set

$$a(x,\xi) = \sum a_{\omega-j}(x,\xi)\theta(\varepsilon_j \xi);$$

the ε_j are chosen so that

$$\left|(1+|\xi|)^{-\mathrm{Re}(\omega)+j+1+|\beta|} D_x^\alpha D_\xi^\beta [a_{\omega-j}(x,\xi)\theta(\varepsilon_j \xi)]\right| \leq 2^{-j}$$

for $|\alpha|$, $|\beta| \leq j$ and (x,ξ) in $K_j \times R^\nu$.

The existence part of Theorem 4 follows immediately from Lemma 2; for f in $C_c^\infty(U)$ and x in U, we put

$$Af(x) = (2\pi)^{-\nu} \int e^{ix\xi} a(x,\xi) \hat{f}(\xi)\, d\xi .$$

To state the continuity properties of these "localized" ψdo's we introduce, for an open U in R^ν:

$H_c^s(U)$ = all F in $H^s(R^\nu)$ supported by a compact subset of U,

$H_{loc}^s(U)$ = all linear maps F: $C_c^\infty(U) \longrightarrow \mathbb{C}$ such that φF is in H^s

for all φ in $C_c^\infty(U)$. Then from Theorem 1 we deduce immediately

<u>Theorem 5</u>. If A is a ψdo_ω in U, then A maps $H_c^s(U)$ into $H_{loc}^{s-Re(\omega)}(U)$.

<u>Proof</u>. Given u in $H_c^s(U)$, choose ψ in $C_c^\infty(U)$ such that $\psi u = u$. Then for any φ in $C_c^\infty(U)$, $\varphi Au = \varphi A\psi u \in H^{s-Re(\omega)}(R^\nu)$, since $\varphi A\psi$ is in $Op(SH^\omega)$.

As an application, we prove the standard result on local regularity for elliptic operators.

<u>Definition 9</u>. Let A be a ψdo_ω in U. A is <u>elliptic</u> (of degree ω) at x if $|\sigma_\omega(A)(x,\xi)| > 0$ for all large ξ. A is elliptic in U if this holds for each x in U.

If A is a matrix (A_{jk}) of ψdo$_\omega$'s in U, then A is elliptic at x if the matrix $(\sigma_\omega(A_{jk})(x,\xi))$ is invertible for all large ξ.

<u>Lemma 3</u>. If A is an elliptic ψdo$_\omega$ in U, then there is a ψdo$_{-\omega}$ B in U such that $\sigma(B)\circ\sigma(A) = 1$. The same holds for elliptic systems.

<u>Proof</u>. We define a sequence $b_{-\omega}, b_{-\omega-1}, \ldots$ by the formula

$$(\Sigma\, b_{-\omega})\circ\sigma(A) = 1.$$

Collecting terms by degree of homogeneity, we find

$$b_{-\omega}\, a_\omega = 1$$

$$b_{-\omega-j}\, a_\omega + \sum_{k>j} (D_\xi^\alpha\, b_{-\omega-k})\,(iD_x)^\alpha\, a_{\omega-m}/\alpha! = 0, \quad j > 0,$$

where the sum is for $k+\alpha+m = j$. Since a_ω is C^∞, homogeneous, and non-vanishing for $|\xi| \geq C$, these formulas define homogeneous C^∞ functions $b_{-\omega-j}$ for $|\xi| \geq C$.

We extend them to all real ξ in any C^∞ way; then by Lemma 2, there is a $\psi do_{-\omega}$ B such that $\sigma(B) = \Sigma\, b_{-\omega-j}$, and Lemma 3 is proved.

When A is an elliptic system, $A = (A_{jk})$, then a_ω is an invertible matrix $(a_{\omega,jk})$, and $b_{-\omega} = a_\omega^{-1}$; the other $b_{-\omega-j}$ are defined by the same formula as before, and the proof of Lemma 3 is completed in the same way.

Theorem 6. Let A be a ψdo_ω in U, and elliptic in $V \subset U$. If u is in $H_c^k(U)$ for some k, and Au is in $H_{loc}^s(V)$, then u is in $H_{loc}^{s+Re(\omega)}(V)$.

Proof. Suppose Au is in $H_{loc}^s(V)$, and φ is given in $C_c^\infty(V)$. Choose ψ in $C_c^\infty(V)$ such that $\psi = 1$ in a neighborhood of the support of φ. By Lemma 3, we have a $\psi do_{-\omega}$ B in V such that $\sigma(B) \circ \sigma(A) = 1$. Writing

$$\varphi u = \varphi B \psi Au + Su \quad ,$$

we note that ψAu is in $H_c^s(v)$ by assumption, hence by Theorem 5 $\varphi B \psi Au$ is in $H^{s+Re(\omega)}$. As for Su, we observe that on a neighborhood of the support of φ, $\sigma(A) = \sigma(\psi A)$, hence, in that neighborhood,

$\sigma(B\psi A) = \sigma(B) \circ \sigma(A) = 1$, hence $\sigma(\varphi B\psi A) = \varphi$, and $\sigma(S) = \sigma(\varphi B\psi A - \varphi) = 0$. It follows that S is a $\psi do_{\omega - J}$ for every J, hence **D** maps u into $H^{s+\mathrm{Re}(\omega)}_{loc}$, and the proof is complete.

Uptto this point we have estimated the remainders in our asymptotic expansions by imposing conditions on the Fourier transform kernels. Sometimes it is more convenient to impose conditions on the operators themselves, as in Kohn and Nirenberg [1]. The final result in this chapter shows that these two methods are equivalent, at least for "localized" ψdo's.

Theorem 7. A is a ψdo_ω in U iff, for each φ and ψ in $C^\infty_c(U)$, there are functions $a_{\omega-j}$ in $S^{\mathrm{Re}(\omega)-j}$, homogeneous of degree $\omega-j$ for $|\xi| \geq 1$, such that for each J,

$$\varphi A\psi - \sum_{j<J} \mathrm{Op}(a_{\omega-j}) \text{ has order } \leq \mathrm{Re}(\omega)-J. \tag{2}$$

Proof. The "only if" part is trivial. If A is a ψdo_ω in U, then $\varphi A\psi = \mathrm{Op}(a)$ with $a \sim \sum a_{\omega-j}$, so $a - \sum_{j<J} a_{\omega-j}$ is in $S^{\mathrm{Re}(\omega)-J}$ and $\mathrm{Op}(a - \sum_{j<J} a_{\omega-j})$ has order $\leq \mathrm{Re}(\omega)-J$, by Theorem II.1.

Conversely, suppose we have functions $a_{\omega-j}$ such that (2) holds. Choose $a \sim \sum a_{\omega-j}$ as in Lemma 2, and choose φ' and ψ' in $C_c^\infty(R^\nu)$ such that $\varphi'\varphi = \varphi$, $\psi\psi' = \psi$. Then $\varphi A\psi$ - Op(a) has order $-\infty$, by (2); multiplying left and right, we find that

$$B = \varphi A\psi - \varphi' \mathrm{Op}(a)\psi' \qquad (3)$$

has order $-\infty$. By Theorem I.3, B has a C^∞ kernel K(x,y), and K has compact support in $R^{2\nu}$ because of the factors $\varphi, \psi, \varphi', \psi'$ in (3). Hence the Fourier transform $\hat{K}(x,\xi)$ of K with respect to y is in the class $S^{-\infty}$. It follows from (2) that $\varphi A\psi = \varphi'\mathrm{Op}(a)\psi' + \mathrm{Op}(\hat{K})$ is in $\mathrm{Op}(SH^{\omega})$, and Theorem 7 is proved.

R. Seeley

III. Kernels of ψdo's.

In this chapter we exhibit the form taken by the kernel of a ψdo. This result shows the connection between classical singular integral operators and ψdo's, and leads to a simple proof of the invariance of ψdo's under diffeomorphisms of R^{ν}.

Definition 1. (i) Let ω be a complex number, $\omega \neq 0,1,2,\ldots$, $Re(\omega) > -\nu$. A function f is a C^{∞} pseudo-homogeneous function of degree ω (ψhf_{ω}) iff, outside the origin, f is a C^{∞} function homogeneous of degree ω.

(ii) If $\omega = 0,1,2,\ldots$, then f is a ψhf_{ω} iff $f(x) = P(x)\log|x| + g(x)$, where P is a polynomial of degree ω and g is C^{∞}, homogeneous of degree ω, and $\int_{|x|=1} x^{\alpha}g(x) = 0$ for $|\alpha| = \omega$.

Remarks. In case (ii) we have separated out the polynomial part of f and multiplied it by $\log|x|$. The Newtonian potential in R^2, which is $C \log|x|$, suggests that these log terms are necessary.

A more thorough discussion of C^∞ homogeneous distributions and ψhf's is given by P. Kree [1].

<u>Definition 2</u>. Let $U \subset R^\nu$ be open, and let $K(x,x-y)$ be defined for x,y in U, $x \neq y$. Let $Re(\omega) > -\nu$. Then K is a pseudo-homogeneous kernel of degree ω (ψhk_ω) iff there are functions $K_{\omega+j}(x,z)$, C^∞ for $z \neq 0$ and pseudo-homogeneous in z of degree $\omega+j$, such that

$$K(x,x-y) - \sum_{j<J} K_{\omega+j}(x,x-y)$$

is of class C^k for $k < Re(\omega) + J$.

Notice that a ψhf of degree μ, $Re(\mu) > 0$, is a function of class C^k for all $k < Re(\mu)$, so the definition is reasonable.

<u>Theorem 1</u>. Let $Re(\omega) < 0$. Then $A: C_c^\infty(U) \longrightarrow C^\infty(U)$ is a ψdo_ω iff

$$Af(x) = \int_U K(x,x-y)f(y)dy$$

with K a $\psi hk_{-\omega-\nu}$.

We could drop the restriction $Re(\omega) < 0$ by allowing distributional kernels. This would require a little more background on homogeneous distributions; see Kree [1].

There are two simple general principles that underlie Theorem 1:

(i) If K is homogeneous of degree s, then $\hat{K}$ is homogeneous of degree $-s-\nu$. This is easy to check simply by writing

$$\hat{K}(t\xi) = \int e^{ixt\xi} K(x)dx ,$$

and setting $xt = y$. However, the integral is not absolutely convergent for any value of s, and not even conditionally convergent for most values of s. This difficulty has to be circumvented in the proof.

(ii) The behavior of $\hat{K}$ near ∞ is closely related to the behavior of K near zero. The proof exploits this to circumbent the convergence problem in (i), and relates the singularities of K near zero to the behavior of $\hat{K}$ at ∞.

The following three lemmas work this out.

Lemma 1. Suppose K is a C^∞ ψhf of degree s, $Re(s) > -\nu$, and ψ is a C^∞ function, $\psi(z) = 1$ for $|z| < \varepsilon$ and $\psi(z) = 0$ for $|z| > R$. Then there is a unique function $a_{-s-\nu}(\xi)$, homogeneous of degree $-s-\nu$ and C^∞ for $\xi \neq 0$, such that for each α and β

$$\xi^\alpha D^\beta [a_{-s-\nu}(\xi) - \widehat{K\psi}(\xi)$$

is bounded in $|\xi| \geq 1$. For $\xi \neq 0$, $a_{-s-\nu}$ is given by

$$a_{-s-\nu}(\xi) = \lim_{t \to +\infty} t^{s+\nu} \widehat{K\psi}(t\xi) \tag{1}$$

Proof. Set

$$A(\xi) = \widehat{K\psi}(\xi) \tag{2}$$

which is C^∞. We cannot study $\lim t^{s+\nu} a(\xi t)$ directly, as (1) suggests, because of convergence problems. Instead, we look at

$$b(t,\xi) = a(\xi) - t^{s+\nu}a(t\xi) \tag{3}$$

$$= \int e^{-iz\xi}\, K(z)\, [\psi(z) - \psi(z/t)]dz.$$

For $|\alpha| > \mathrm{Re}(s)+\nu$ and $t > 1$ we get

$$\xi^\alpha b(t,\xi) = \int_{|z|>\varepsilon} e^{-iz\xi}\, D_z^\alpha \{K(z)[\psi(z) - \psi(z/t)]\}dz$$

$$= \int_{|z|>\varepsilon} e^{-iz\xi}(D^\alpha K)[\psi(z) - \psi(z/t)]dz$$

$$+ \sum_{\beta<\alpha} C_\beta \int_{|z|>\varepsilon} e^{-iz\xi}(D^\beta K)[\psi_{\alpha-\beta}(z) - t^{|\beta|-|\alpha|}\psi_{\alpha-\beta}(z/t)]dz$$

$$\longrightarrow \int_{|z|>\varepsilon} e^{-iz\xi}(D^\alpha K)(\psi - 1)dz$$

$$+ \sum_{\beta<\alpha} C_\beta \int e^{-iz\xi}(D^\beta K)\, \psi_{\alpha-\beta}\, dz \tag{4}$$

in other words,

$$\xi^\alpha b(t,\xi) \longrightarrow \int e^{-iz\xi} D^\alpha[K(\psi - 1)]dz \tag{5}$$

$$= b_\alpha(\xi). \tag{6}$$

By way of explanation: We can take $\int_{|z|>\varepsilon}$ since $\psi(z)$ and $\psi(z/t)$ both = 1 for $|z| < \varepsilon$; in the second equality we set $\psi_{\alpha-\beta} = D^{\alpha-\beta}\psi$, and use Leibniz' rule; the first term in the limit (4) exists since $|D^\alpha K(z)| \leq \text{const.}\ |z|^{Re(s)-|\alpha|}$, and $Re(s) - |\alpha| < -\nu$; the other limits in (4) exist since ψ is constant for $|z| > R$, whence $\psi_{\alpha-\beta}(z/t) = 0$ for $|z/t| > R$, whence

$$t^{|\beta|-|\alpha|} \left| \int_{|z|>\varepsilon} e^{-iz\xi}(D^\beta K)\psi_{\alpha-\beta}(z/t)dz \right|$$

$$\leq \text{constant} \cdot t^{|\beta|-|\alpha|} \int_\varepsilon^{tR} r^{Re(s)-|\beta|+\nu-1} dr$$

$$= c_1 t^{Re(s)-|\alpha|+\nu} + c_2 t^{|\beta|-|\alpha|} \longrightarrow 0 \quad \text{as } t \longrightarrow +\infty.$$

Formula (5) follows from (4) by Leibniz' rule, and (6) defines b_α. Clearly b_α is bounded, and $\xi^\beta b_\alpha = b_{\alpha+\beta}$, so there is a unique function $b_o(\xi)$ such that

$$b_\alpha(\xi) = \xi^\alpha b_o(\xi), \quad \xi \neq 0. \tag{7}$$

In fact, if $2k > \mathrm{Re}(s) + \nu$, and constants C_α are chosen so that $|\xi|^{2k} = \sum_{|\alpha|=2k} C_\alpha \xi^\alpha$, then $b_o = |\xi|^{-2k} \sum C_\alpha b_\alpha$.

Now we set $a_{-s-\nu} = a - b_o$; then it follows from (2),(3),(5), and (7), that

$$\xi^\alpha a_{-s-\nu} = \xi^\alpha (a - b_o) = \lim_{t \to +\infty} \xi^\alpha t^{s+\nu} a(t\xi), \tag{8}$$

which proves formula (1). The homogeneity of $a_{-s-\nu}$ follows from (1).

We have proved the boundedness of $\xi^\alpha D^\beta [a_{-s-\nu} - a]$ for $\beta = 0$; for general β

we apply the same argument to $\xi^\alpha D^\beta b(t,\xi)$ in place of $\xi^\alpha b$, taking initially $|\alpha| > Re(s) + \nu + |\beta|$.

To show that $a_{-s-\nu}$ is independent of ψ, notice that if $\psi_1 = 0$ near ∞ and $\psi_1 = 1$ near 0, then $K(\psi - \psi_1)$ is C^∞ with compact support, so $\lim t^{-s-\nu} \int K(z)[\psi(z) - \psi_1(z)]\, e^{-iz\xi} dz = 0$.

Q.E.D.

Lemma 1 shows how to pass from a "pseudo homogeneous" kernel K_s to a symbol $a_{-s-\nu}$; we recover $a_{-s-\nu}$ indirectly from the function $\xi^\alpha a_{-s-\nu}$ for sufficiently large α. The passage in the other direction, from a symbol a_s to a kernel $K_{-s-\nu}$, is similar, except that now we work through the <u>derivatives</u> $D^\alpha K_{-s-\nu}$ for sufficiently large α.

<u>Lemma 2.</u> Let k be an integer ≥ 0, and $Re(\omega) - k > -\nu$. A function f is a ψhf_ω plus a polynomial of degree $< k$ iff $D^\alpha f$ is a $\psi hf_{\omega-k}$ for all $|\alpha| = k$.

<u>Proof.</u> It is clear from Definition 1 that $D^\alpha f$ is a $\psi hf_{\omega-k}$. For

the converse, it suffices to take $k=1$. If $\mathrm{Re}(\omega)>0$ we have

$f(x) = \int_0^x df + c$, and the Lemma follows easily. If $\mathrm{Re}(\omega)<0$ we find that

$f(x) - \int_\infty^x df = f(x) - \int_\infty^1 \Sigma x_j (df)_j(tx)dt$ is constant, so $f(x) = \int_\infty^x df + c$.

Finally, if $\mathrm{Re}(\omega) = 0$, we can write

$$df(x) = a|x|^{\omega-2}\Sigma x_j dx_j + \Sigma G_j dx_j , \tag{9}$$

where the G_j are homogeneous of degree $\omega - 1$ and

$$\int_{|x|=1} \Sigma G_j x_j = 0; \tag{10}$$

we simply choose the constant $\underline{a}$ so that $a \cdot \int_{|x|=1} \Sigma x_j^2 = \int_{|x|=1} x_j x_j \partial f/\partial x_j$,

and define the G_j by (9). We have

$$a|x|^{\omega-2}\Sigma x_j dx_j = \begin{cases} d(\frac{a}{\omega}|x|^{\omega}), & \omega \neq 0 \\ d(a \log|x|), & \omega = 0. \end{cases} \tag{11}$$

Hence $G = \sum G_j dx_j$ is exact, since df is. Choose a point $\bar{x}$ such that $|\bar{x}| = 1$ and

$$\sum \bar{x}_j G_j(\bar{x}) = 0 \tag{12}$$

which is possible, by (10), and set $g(x) = \int_{\bar{x}}^{x} G$. Then $dg = G$ for $x \neq 0$, and g is homogeneous of degree ω. To prove the homogeneity, note that $\int_{\bar{x}}^{r\bar{x}} G = 0$ for all $r > 0$, by (12), so

$$g(x) = \int_{\bar{x}}^{|x|\bar{x}} G + \int_{|x|\bar{x}}^{x} G = 0 + |x|^{\omega} \int_{\bar{x}}^{x/|x|} G = |x|^{\omega} g(x/|x|) \ .$$

<u>Remark.</u> If df depends C^∞ on some parameters, then so does f.

<u>Lemma 3.</u> Let $a(\xi)$ be C^∞ and homogeneous of degree s, $Re(s) < -\nu$, and let θ be C^∞, $\theta(\xi) = 1$ for ξ large, $\theta = 0$ for ξ small. Then

$$K(z) = \int e^{iz\xi} a(\xi)\theta(\xi) d\xi$$

is continuous, and there is a unique ψhf K_o of degree $-s-\upsilon$ such that $K(z) - K_o(z)$ is C^∞.

<u>Proof.</u> Since $a\theta \varepsilon L^1$, K is continuous; and differentiating under the integral, we find in the same way, that $D^\alpha K$ is continuous for $\mathrm{Re}(s) + |\alpha| < -\upsilon$. For $z \neq 0$, we integrate by parts to obtain

$$K(z) = \int e^{iz\xi} |z|^{-2k} \triangle_\xi^k (\theta a) d\xi ,$$

and conclude that $D^\alpha K$ is continuous for $|\alpha| < -\upsilon - \mathrm{Re}(s) + 2k$. Since k can be taken arbitrarily large, K is C^∞ for $z \neq 0$.

Now set $L(t,z) = K(z) - t^{s+\upsilon} K(tz)$, $t > 0$. we expect $\lim_{t \to 0+} t^{s+\upsilon} K(tz) = K_o(z)$, and hence $\lim_{t \to 0+} L(t,z)$ should equal $K(z) - K_o(z)$. Actually, we can only take $\lim_{t \to 0+}$ for certain <u>derivatives</u> of $L(t,z)$; then we apply the preceding lemma.

We have

$$L(t,z) = K(z) - t^{s+\upsilon} K(tz) = \int_{|\xi| \leq c} e^{iz\xi} [\theta(\xi) - \theta(\xi/t)] a(\xi) d\xi .$$

If $|\alpha| > -\nu-\mathrm{Re}(s)$, we can take the limit

$$\lim_{t\to 0+} D^{\alpha}L(t,z) = \int_{|\xi|\le c} e^{iz\xi}\,\xi^{\alpha}[\theta(\xi) - 1]a(\xi)d\xi$$

$$= L_{\alpha}(z) \qquad \text{(definition of } L_{\alpha}\text{)}$$

The L_{α} satisfy $D^{\beta}L_{\alpha} = L_{\alpha+\beta}$, so there is a C^{∞} function L_o such that $D^{\alpha}L_o = L_{\alpha}$, $|\alpha| > -\nu-\mathrm{Re}(s)$. To prove that $K-L_o$ differs from a pseudo-homogeneous function by a polynomial, notice that for $|\alpha| > -\nu-\mathrm{Re}(s)$ we have

$$D^{\alpha}(K-L_o)(z) = D^{\alpha}K(z) - L_{\alpha}(z) = \lim_{t\to 0+} t^{s+\nu+|\alpha|}\, K(tz).$$

This limit is, by its very form, homogeneous of degree $-s-\nu-|\alpha|$, so Lemma 2 above shows that $K-L_o$ differs by a polynomial from some $\psi hf, K_o$. This K_o satisfies all the conditions of the theorem; it is unique, for if K_1 has the same properties as K_o, then $K_1 - K_o$ is a ψhf and C^{∞} at the origin, hence $K_1 - K_o = 0$.

Remark. If a depends smoothly on some parameters, it follows from our construction that K_o depends smoothly on those parameters.

Proof of Theorem 1. Suppose A is a ψdo in U. It suffices to prove that for each φ and ψ in $C_c^\infty(U)$, φAψ has a ψhk. Let φAψ = Op(a) with a in $SH^{(\omega)}$. We may assume Re(ω) < -ν; for otherwise, we can replace a(x,ξ) by $b=a(x,\xi)(1+|\xi|^2)^{-k}$, obtain a kernel K(x,x-y) for Op(b), and thus find the kernel $(1+\Delta_y)^k K(x,x-y)$ for Op(a).

Thus we have $a \sim \sum \theta(\xi) a_{\omega-j}(x,\xi)$, where θ is a C^∞ function as in Lemma 3, and $a_{\omega-j}$ is homogeneous of degree ω-j. Since a is in $S^{Re(\omega)}$ with Re(ω) < -ν, φAψ = Op(a) has the continuous kernel

$$K(x,x-y) = (2\pi)^{-\nu} \int e^{i(x-y)\xi}\, a(x,\xi)d\xi ,$$

while $Op(\theta a_{\omega-j})$ has a similar kernel, and $\varphi A\psi - \sum_{j<J} Op(\theta a_{\omega-j})$ has the kernel

$$K_J(x,x-y) = (2\pi)^{-\nu} \int e^{i(x-y)\xi} \,[a(x,\xi) - \sum_{j<J} \theta a_{\omega-j}]\, d\xi .$$

Since $a - \sum_{j<J} \theta a_{\omega-j}$ is in $S^{\mathrm{Re}(\omega)-J}$, this last kernel is of class C^k for $k < J - \nu - \mathrm{Re}(\omega)$. Hence it follows from Lemma 3 that $K(x,x-y)$ is a $\psi hk_{-\omega-\nu}$.

To go in the other direction, let $Af(x) = \int K(x,x-y)f(y)dy$ with K a $\psi hk_{-\omega-\nu}$, $\mathrm{Re}(\omega) < 0$. Then the kernel of $\varphi A\psi$ is $K'(x,x-y) = \varphi(x)K(x,x-y)\psi(y)$, and expanding $\psi(y)$ in powers of $x-y$ shows that this, too, is a $\psi hk_{-\omega-\nu}$, $K'(x,z) \sim \Sigma K_{-\omega-\nu+j}(x,z)$. Choose ψ' in $C_c^\infty(R^\nu)$ such that $\psi'(x-y) = 1$ when $\varphi(x)\psi(y) \neq 0$; then $K'(x,x-y)\psi'(x-y) = K'(x,x-y)$. Set

$$a(x,\xi) = \int e^{-iz\xi} K'(x,z)dz ,$$

$$a_{\omega-j}(x,\xi) = \int e^{-iz\xi} K_{-\omega-\nu+j}(x,z)\psi'(z)dz.$$

Then $\varphi A\psi = Op(a)$, and

$$a - \sum_{j<J} a_{\omega-j} = \int e^{-iz\xi}[K' - \sum_{j<J} K_{-\omega-\nu+j}\ \psi'](x,z)dz .$$

The integrand on the right is of class C_c^k for $k < \mathrm{Re}(J-\omega-\nu)$, so

$\xi^\alpha D_\xi^\beta D_x^\gamma [a - \sum_{j<J} a_{\omega-j}]$ is bounded for all β and $|\alpha+\gamma| < \mathrm{Re}(J-\omega-\nu)$. By

Lemma 1, each $a_{\omega-j} \sim a'_{\omega-j}$ where $a'_{\omega-j}$ is homogeneous of degree $\omega-j$

for large ξ; it follows easily that $a \sim \Sigma\, a'_{\omega-j}$ in the sense of

Definition II.3, and Theorem 1 is proved.

As a first consequence of Theorem 1, we deduce a relation between ψdo's and classical singular integral operators: if a is in SH^0, then Op(a) is a classical singular integral operator. To prove this, let $b = a(x,\xi)(1+|\xi|^2)^{-1}$. Then by Theorem 1, Op(b) has a ψ$hk_{2-\nu}$, and writing $A = \mathrm{Op}(b)(1+\Delta)$ we obtain A as a singular integral operator.

As a second consequence, we deduce the effect of coordinate changes. Let $\chi: V \longrightarrow U$ be a C^∞ diffeomorphism of open sets in R^ν. Let A be a ψdo_ω in U, and suppose at first that $\mathrm{Re}(\omega) < 0$. Then

$$Af(\bar{x}) = \int_U K(\bar{x},\bar{x}-\bar{y})f(\bar{y})d\bar{y} = \int_V \tilde{K}(x,x-y)\tilde{f}(y) \int_V \tilde{K}(x,x-y)\tilde{f}(y)dy, \text{ where}$$

$\tilde{K}(x,x-y) = K(\chi(x),\chi(x)-\chi(y))\,|\det \partial\chi/\partial y|$, $\tilde{f}(y) = f(\chi(y))$. If we substitute the Taylor expansions of $\chi(y)$ and $|\det \partial\chi/\partial y|$ about the point x into the ψhk expansion of K, we get a corresponding expansion for the kernel $\tilde{K}$ operating in V.

If $K(\overline{x},\overline{z}) \sim \Sigma K_{-\omega-\nu+j}(\overline{x},\overline{z})$, then $\tilde{K}(x,z) \sim \Sigma \tilde{K}_{-\omega-\nu+j}(x,z)$, where in particular $\tilde{K}_{-\omega-\nu}(x,z) = K_{-\omega-\nu}(\chi(x), J(x)\cdot z)\, v(x)$, with $J(x) = \partial\chi/\partial x$, $v(x) = |\det J(x)|$. Denoting by $\Sigma a_{\omega-j}$ and $\Sigma \tilde{a}'_{\omega-j}$ the symbols of A (acting in U) and $\tilde{A}^*$ (acting in V), we find from Lemma 3 that

$$\tilde{a}_{\omega}(x,\xi) = \lim_{t\to\infty} t^{-\omega} \int e^{-itz\xi}\, \tilde{K}_{-\omega-\nu}(x,z)\psi(z)dz$$

$$= \lim_{t\to\infty} t^{-\omega} \int e^{-itw\overline{\xi}}\, K_{-\omega-\nu}(\chi(x),w)\psi(J^{-1}w)dw\ ,$$

$$\tilde{a}_{\omega}(x,\xi) = a_{\omega}(\chi(x),\overline{\xi})\ , \tag{13}$$

where we have set $\xi = J(x)^t\overline{\xi}$ and $w = J(x)z$, (with $J(x)^t$ = the transpose of $J(x)$.)

We have almost proved:

<u>Theorem 2</u>. Let $\chi: V \longrightarrow U$ be a C^∞ diffeomorphism of open sets in R^ν, and let A be a ψdo_ω in U. Then if $\tilde{f}(x) = f(\chi(x))$, and $\tilde{A}$ is defined by

$\tilde{A}\tilde{f}(x) = Af(\chi(x))$, it follows that $\tilde{A}$ is a ψdo_ω, and the leading terms in the symbols of A and A are related by (13).

Proof. If $Re(\omega) < 0$, the theorem has just been proved. If A is actually a differential operator, the theorem is trivial. The general case is obtained from these two by taking products. Notice that it suffices to consider $\widetilde{\varphi A \psi}$ for φ, ψ in $C_c^\infty(U)$. But $\varphi A \psi = (\varphi A \psi \Lambda^{-2k}) \Lambda^{2k}$; when $2k > Re(\omega)$, the first factor here can be transferred from U to V, as we just proved, and the second factor is simply a differential operator. The relation between the new and old symbols is obtained by observing that it holds for each factor, and that the top order symbol of a product is the product of the top order symbols.

Remarks. There are two other ways to prove the invariance under coordinate changes, one working directly with the Fourier integral representation of Op(a), and another relying on an intrinsic, coordinate-free definition of ψdo's. The first approach has been worked out very neatly by Kuranishi, and can be found in the notes of the lectures of L. Nirenberg [1].

The second approach, due to Hormander [2], is based on an asymptotic expansion of $e^{-i\lambda g}Ae^{i\lambda g}f$, where $dg \neq 0$ on the support of f. The direct proof that such an expansion exists when A is a ψdo (in our sense) is the most difficult and most important part of this development. We can obtain a fairly simple proof of this expansion, based on the results above.

Theorem 3. Let A be a ψdo_ω in U, $\sigma(A) = \Sigma\, a_{\omega-j}$; let $f \in C_c^\infty(U)$, $g \in C^\infty(U)$ and $|dg| \neq 0$ on the support of f. Then as $\lambda \to +\infty$ we have the asymptotic relation

$$[e^{i\lambda g}Ae^{-i\lambda g}f](x) \sim \sum_\alpha \sum_j a^{(\alpha)}_{\omega-j}(x, \lambda dg_x)\, D^\alpha(fe^{i\lambda hx})(x)\ , \tag{14}$$

where $a^{(\alpha)}(x,\xi) = (\partial/\partial\xi)^\alpha a(x,\xi)$, and $g(x) - g(y) = dg_x\cdot(x-y) + h_x(y)$.

Notice that $h_x(y)$ vanishes to second order at y=x; hence evaluating $D^\alpha_y(f(y)e^{i\lambda h_x(y)})$ at y=x yields integer powers λ^k with $k \leq |\alpha|/2$. Since $a^{(\alpha)}_{\omega-j}$ is homogeneous of degree $\omega-j-|\alpha|$, the terms on the right of (14) can be grouped into the powers λ^ω, $\lambda^{\omega-1}$,..., and there are only finitely many involving a given power of λ. Say $S_N(x,\lambda)$ is the sum of all terms involving $\lambda^{\omega-j}$ for $j \leq N$. Then the asymptotic relation in (14) means that for each N, α, and compact $C \subset U$,

$$\lambda^{N-\mathrm{Re}(\omega)} D^{\alpha}[e^{i\lambda g} A e^{-i\lambda g} f(x) - S_N(x,\lambda)]$$

is bounded for x in C and $\lambda \geq 1$. The proof will show that the bound is fixed even as f and g run through bounded sets of C_c^∞ and C^∞, subject to the condition that there is a fixed $\varepsilon > 0$ such that $|dg| \geq \varepsilon$ on the support of f.

By a partition of unity, the proof of Theorem 3 is divided into two cases:

(i) x and the support of f both lie in a compact set C where some derivative of g is bounded away from zero, say $|\partial g/\partial x_1| \geq \varepsilon > 0$, and $|dg_x \cdot (x-y)| > 2\,|h_x(y)|$ for x and y in C.

(ii) the support of f lies in a set C as in (i), but x lies in a compact set C' disjoint from C.

The second case follows easily from the first; we can set g' = g in C, and for y in C' set $g'(y) = dg_{x_0} \cdot (y-x_0)$, where x_0 is a fixed point in C. Then C' is a finite union of compact sets C_n such that in $C \cup C_n$, g' satisfies the conditions in (i). Hence for x in C', $D^\alpha e^{-i\lambda g'} A e^{i\lambda g'} f$ is $O(|\lambda|^{-N})$ for every α and N. Since $e^{i\lambda g'} f = e^{i\lambda g} f$, we find $D^\alpha e^{-i\lambda g} A e^{i\lambda g} f$ to be $O(|\lambda|^{-N})$ for every α and N, and Theorem 3 follows.

Turning to the proof of case (i), we have

$$e^{i\lambda g}Ae^{-i\lambda g}f(x) = \iint e^{i\lambda[g(x)-g(y)]}e^{i(x-y)\xi}a(x,\xi)f(y)dy\,d\xi$$

$$= \iint E\, e^{i\lambda h_x(y)}a(x,\xi)f(y)dy\,d\xi$$

$$= \sum_0^N \iint E(i\lambda h_x)^n\, a\, f\, dyd\xi/n! + R_N/N!,$$

where $E = e^{i(x-y)(\xi+\lambda dg_x)}$ and

$$R_N = \iiint E(i\lambda h_x)^{N+1}(1-t)^N e^{it\lambda h_x} a\, f\, dt\, dy, d\xi/N.$$

Repeating the proof of Lemma II. _ , we have

$\iint E\,(i\lambda h_x)^n\, a\, f\, dyd\xi/n! \sim \Sigma\, a^{(\alpha)}(x,\lambda dg_x)\, D^\alpha(f(i\lambda h_x)^n)(x)/n!$ Since the sum of these terms over n and α gives the same expansion as (14), it remains only to estimate the remainder R_N.

Here we write $h_x(y)^{N+1} = \sum_{|\alpha|=2N+2} h_\alpha(x,y)(x-y)^\alpha$, where h_α is C^∞ in (x,y).

Integration by parts and differentiation under the integral yields

$$R_N = (i\lambda)^{N+1} \Sigma \iiint E(1-t)^N e^{it\lambda h_x} a^{(\alpha)}(x,\xi) h_\alpha f \, dt\,dy\,d\xi \; .$$

When $|\alpha| = 2N+2 > \mathrm{Re}(\omega) + \nu$, we can interchange the order of integration, obtaining

$$R_N = (i\lambda)^{N+1} \Sigma \iint (1-t)^N K_\alpha(x,x-y) e^{i(x-y)\lambda dg_x + it\lambda h_x} h_\alpha f \, dy\,dt \; ,$$

where $K_\alpha(x,x-y)$ is a ψdk of degree $2N+2-\omega-\nu$. By the hypotheses in case (i), we can make the change of variable

$$z_1 = dg_x \cdot (x-y) + t\,h_x(y)\, , \; z_2 = x_2 - y_2, \ldots, z_\nu = x_\nu - y_\nu$$

and obtain

$$R_N = (i\lambda)^{N+1} \sum_\alpha \iint K'_\alpha(x,t,z) e^{i\lambda z_1} \, dz\,dt \; ,$$

where $(x,t) \longrightarrow K'_\alpha(x,t,z)$ is a C^∞ map of $C \times [0,1]$ into pseudo-differential kernels of degree $2N+2-\omega-\nu$, with compact support in the z variable. It follows from Lemma III.1 that R_N and its derivatives are $O(|\lambda|^{Re(\omega)-N-1})$, and the proof of Theorem 3 is complete.

IV. Ψdo's on compact manifolds.

Let M be a C^∞ manifold, and A a map: $C_c^\infty(M) \longrightarrow C^\infty(M)$. Then for each chart χ mapping a domain $V \subset M$ onto an open set U_χ in R^ν, we have in a natural way a map $A_\chi : C_c^\infty(U_\chi) \longrightarrow C^\infty(U_\chi)$.

Definition 1. A is a Ψdo_ω on M iff every such A_χ is a Ψdo_ω in U_χ. The space of these operators is denoted by $\Psi do_\omega(M)$. If A is a Ψdo_ω and x is in the domain of a chart χ, we define the top-order symbol of A by

$$\sigma_\omega(A)(\Sigma\, \xi_j d\chi^j(x)) = a_\omega(\chi(x), \xi),$$

where $\xi = (\xi_1, \dots, \xi_\nu) \neq 0$, $d\chi^j(x)$ is the basis of the cotangent plane at x associated with the chart χ, and a_ω is the leading term in $\sigma(A_\chi) = \Sigma a_{\omega-j}$.

Notice that every pair of points x,y in M lies in the domain of some chart χ, so the maps A_χ determine A completely.

According to the transformation law for a_ω given in Theorem III.2, $\sigma_\omega(A)$ is a well-defined function on the bundle of non-zero cotangent vectors. (We denote this bundle by $T'(M)$).

The symbol can be given a more intrinsic definition, using the asymptotic expansion in Lemma II.1. Given a non-zero cotangent vector ξ_x in T'_x, choose a function g with $dg_x = \xi_x$. Then g can be considered as the first component of some diffeomorphism of a neighborhood U of x into R^ν. Pick φ in $C_c^\infty(V)$ with $\varphi(x) = 1$. From Lemma II.1 we find that $\sigma_\omega(A)(\xi_x)$ is the coefficient of λ^ω in the expansion of $[e^{-i\lambda g} A\varphi e^{i\lambda g}](x)$.

When A is a differential operator of order ≤ 1, this coefficient can be determined in a simpler way. Take $dg_x = \xi_x$ and $g(0) = 0$, g in $C_c^\infty(M)$. Then from the expression given in the previous paragraph it follows easily that $\sigma_1(A)(\xi_x) = iAg(x)$.

For the rest of this chapter we suppose that M is compact. For convenience, we suppose that a fixed C^∞ measure v is given on M. We take the Sobolef space $H^s(M)$ to be the completion of $C^\infty(M)$ under some norm defined by a partition of unity subordinate to a covering of M by domains of charts. Then H^s is a topological "Hilbertable" space, but is not endowed with any particular norm except in the case $s = 0$, where we take $H^o(M)$ isometric with $L^2(v)$. The inner product $\int fgdv$ extends to a pairing of H^{-s} and H^s which realizes each of these as the dual of the other, and the pairing $\int f\bar{g}dv$ gives a conjugate pairing of H^s and H^{-s}. (To justify all this, see Chapter I, particularly Theorem 1, and Corollaries 1 and 2 of Theorem 5.)

Theorem 1. Let A be a ψdo_ω and B a ψdo_μ. Then

(i) A has order $\leq Re(\omega)$, i.e., A is continuous from H^s to $H^{s-Re(\omega)}$; and A has order $\leq Re(\omega) - 1$ if $\sigma_\omega(A) = 0$.

(ii) If $\omega = \mu$, $A+B$ is a ψdo_ω, and $\sigma_\omega(A+B) = \sigma_\omega(A) + \sigma_\omega(B)$.

(iii) AB is a $\psi do_{\omega+\mu}$ and $\sigma_{\omega+\mu}(AB) = \sigma_\omega(A)\sigma_\mu(B)$.

(iv) There are ψdo_ω's A^* and A' such that

$$\int(Af)g = \int f(A'g), \quad \int(Af)\bar{g} = \int f\overline{(A^*g)},$$

and $\sigma_\omega(A') = \sigma_\omega(A)$, $\sigma_\omega(A^*) = \overline{\sigma_\omega(A)}$.

(v) If $\varphi\psi = 0$, φ and ψ in $C^\infty(M)$, then $\varphi A\psi$ has order $-\infty$, and has a C^∞ kernel.

Proof. Parts (i) and (ii) follow directly from Definition 1 and the corresponding properties of ψdo's in an open set U of R^ν. For part (iii) notice there is a C^∞ partition of unity $\Sigma\varphi_j = 1$ such that any four φ_i, φ_j, φ_k, φ_ℓ have their supports all in the domain of a single chart χ. Since $AB = \Sigma\ \varphi_i A\varphi_j\varphi_k B\varphi_\ell$, we are reduced to the study of the individual terms in this sum, each of which can be analyzed in local coordinates. From Theorem II.4, the operator

$$C_\chi = (\varphi_i A\ \varphi_j\varphi_k\ B\ \varphi_\ell)_\chi = (\varphi_i A\ \varphi_j)_\chi(\varphi_k B\ \varphi_\ell)_\chi$$

is a $\psi do_{\omega+\mu}$ and

$$\sigma_{\omega+\mu}(C_\chi)(x,\xi) = \sigma_\omega(A)(x,\xi)\sigma_\omega(B)(x,\xi)\varphi_i\varphi_j\varphi_k\varphi_\ell(\chi^{-1}(x)).$$

From Theorem III.2, C is a $\psi do_{\omega+\mu}$ and

$$\sigma_{\omega+\mu}(C) = \varphi_i\varphi_j\varphi_k\varphi_\ell\ \sigma_\omega(A)\sigma_\mu(B).$$

Now (iii) follows by addition over $i,j,k\ell$.

The proof of (iv) is similar. For (v), we reduce the question to R^ν in the usual way, and there we observe that $\varphi D^{\alpha}\psi \equiv 0$ for all α, so that all the terms in the expansion of $\sigma(\varphi A\psi)$ vanish identically, hence $\varphi A\psi$ has order $-\infty$. The C^∞ kernel is obtained then either from Theorem III.1 or from Theorem I.3.

Next we derive the basic facts about elliptic operators. Since the main geometric applications of these results involve vector bundles, we will state them in that context. Definition 1 has to be altered only slightly to cover this case. If E is a p-dimensional vector bundle over M, then $E = \bigcup_{x\varepsilon M} E_x$ is a disjoint union of p-dimensional vector spaces E_x, one for each point of M, together with certain "trivializing" charts we will describe. If $U \subset M$, then $E_U = \bigcup_{x\varepsilon U} E_x$ is the restriction of E to U. A <u>chart</u> for E is a map τ of some E_{U_τ} onto $V_\tau \times \mathbb{C}^p$, where $V \subset R^\nu$, satisfying

(i) For each x in U_τ, there is a $\bar{\tau}(x)$ in V_τ such that

$\tau(E_x) = \bar{\tau}(x) \times C^p$,

(ii) $\bar{\tau}$ is a diffeomorphism of U_τ onto V_τ ,

(iii) The induced map τ_x of E_x onto C^p is an isomorphism of vector spaces.

We suppose that there is a maximal collection of these charts τ such that the U_τ cover M, and any two τ, τ' are compatible, i.e. $\bar{\tau}' \bar{\tau}^{-1}$ is a diffeomorphism, and the map

$$x \longrightarrow \tau'_x \tau_x^{-1}$$

is a C^∞ map of $U_\tau \cap U_{\tau'}$ into invertible transformation on $\mathbb{C}^p$.

These charts make E a C^∞ manifold. A <u>section</u> of E is a map $f: M \longrightarrow E$ such that $f(x) \in E_x$ for each x. The C^∞ sections are denoted $C^\infty(E)$. Given a chart τ over U, and taking $\varepsilon_1, \ldots, \varepsilon_p$ to be the usual basis of $\mathbb{C}^p$, we get sections τ_j of E_U defined by $\tau_j(x) = \tau_x^{-1}(\varepsilon_j)$. Every C^∞ section of E_U is then a linear combination $\Sigma f_j\tau_j$, where f_j is in $C^\infty(U)$. Thus $C^\infty(E_U)$ is isomorphic to $C^\infty(U)^p$, where p is the dimension of E.

We define $H^s(E)$ as the completion of $C^\infty(E)$ under any appropriate norm constructed by using partitions of unity. We suppose that each fibre E_x carries a Hermitian inner product $(\ ,\ )_x$, such that $x \longrightarrow (f(x),g(x))_x$ is a C^∞ function for every f and g in $C^\infty(E)$; with this inner product we define the norm in $H^0(E)$ to be $(f,g) = \int (f(x),g(x))_x dv(x)$. The pairing (f,g) gives an anti-isomorphism between H^{-s} and the dual of H^s.

Now suppose we are given a linear map $A: C^\infty(E) \longrightarrow C^\infty(F)$, where E and F are vector bundles over M, of dimensions p and q respectively. Given trivializing charts τ and τ' of E_U and F_U respectively, we obtain isomorphisms of $C^\infty(E_U)$ with $C^\infty(U)^p$ and of $C^\infty(F_U)$ with $C^\infty(U)^q$;

<u>Definition 2</u>. A is a ψdo_ω iff each such $A_{\tau\tau'}$ is a $p \times q$ matrix of ψdo_ω's on U. The space of such operators is denoted $\psi do_\omega(E,F)$. If $A_{\tau\tau'} = (A_{jk})$, and if ξ_x is a non-zero cotangent vector, then $\sigma_\omega(A)(\xi_x)$ is the linear map of E_x into F_x defined by

$$\sigma_\omega(A)(\xi_x)\ \Sigma\, f_j\tau_j(x) = \sum_{j,k} [\sigma_\omega(A_{jk})(\xi_x)f_j]\ \tau_k'(x),$$

where $\{\tau_j\}$ and $\{\tau_k'\}$ are the local bases of $C^\infty(E_U)$ and $C^\infty(F_U)$ described above.

The symbol can be defined more intrinsically as follows: Given $\xi_x \neq 0$ and e_x in E_x, choose g such that $dg_x = \xi_x$, choose a neighborhood V of x such that g is a local coordinate in V, and let φ be a section in $C^\infty(\mathbf{E}_V)$ with $\varphi(x) = e_x$. Then $\sigma_\omega(A)(\xi_x)e_x$ is the coefficient of λ^ω in the expansion of $[e^{-i\lambda g}Ae^{i\lambda g}\varphi](x)$. From Lemma II.1 it is easy to see

(i) this first term depends only on $\varphi(x)$ and dg_x

(ii) this agrees with the definition of $\sigma_\omega(A)$ in Definition 2.

Hence $\sigma_\omega(A)$ is well-defined.

When A is a differential operator of order ≤ 1, we get the following simple description: if g in $C_c^\infty(M)$ and φ in $C_c^\infty(E)$ are chosen so that $g(x) = 0$, $dg_x = \xi_x$, and $\varphi(x) = e_x$, then $\sigma_1(A)(\xi_x)e_x = (A\, g\, \varphi)(x)$.

Theorem 1 carries over to vector bundles in a routine way, using the trivilizations $A_{TT'}$ instead of A_X. [(Note that the transpose $A': C^\infty(F') \longrightarrow C^\infty(E')$ acts on the dual bundles to E and F)].

A ψdo$_\omega$ $A: C^\infty(E) \longrightarrow C^\infty(F)$ is <u>elliptic</u> (of degree ω) iff $\sigma_\omega(A)(\xi_x)$ is an isomorphism of E_x onto F_x for every $\xi_x \neq 0$. When A acts on $C^\infty(M)$, this says simply that $\sigma_\omega(A)(\xi_x) \neq 0$ for $\xi_x \neq 0$. The first results on elliptic operators depend on the existence of a parametrix.

<u>Lemma 1.</u> If A is an elliptic ψdo$_\omega$: $C^\infty(E) \longrightarrow C^\infty(F)$, then there is a ψdo$_{-\omega}$ $B: C^\infty(F) \longrightarrow C^\infty(E)$ such that $BA - I_E$ and $AB - I_F$ are of degree $-\infty$. (Here I_E and I_F denote the identity operators in $C^\infty(E)$ and $C^\infty(F)$.)

<u>Proof.</u> Cover M with finitely many open sets U_j such that E_{U_j} and F_{U_j} possess trivializing charts. Choose a partition of unity $\Sigma\, \varphi_j = 1$ with φ_j in $C_c^\infty(U_j)$, and choose functions ψ_j and θ_j in $C_c^\infty(U_j)$ such that $\psi_j\varphi_j = \varphi_j$ and $\theta_j\psi_j = \psi_j$. By Lemma II.3, we have a B_j such that $C_j = (\varphi_j B_j \psi_j)(\theta_j A \theta_j) - \varphi_j$ has degree $-\infty$; Lemma II.3 provides B_j in one coordinate system, and Theorem III.2

shows that $\varphi_j B_j \psi_j$ is a $\psi do_{-\omega}$ on M. Setting $B = \Sigma\, \varphi_j B_j \psi_j$, we get

$$BA - I_E = \Sigma\, C_j + \Sigma\, \varphi_j B_j \psi_j A(1-\theta_j).$$

Here C_j has degree $-\infty$, and $\psi_j(1-\theta_j) = 0$, so by Theorem 1(v) $BA - I_E$ has degree $-\infty$.

So far we have a left parametrix for A. We can then obtain a right parametrix $\tilde{B}$ as the adjoint of a left parametrix of A^*. Finally, to show that the left parametrix B is also a right parametrix of A, we observe that $BA\check{B}-B$ and $BA\tilde{B}-\tilde{B}$ both have degree $-\infty$, so $\tilde{B} - B$ has degree $-\infty$, and $AB - I$ has degree $-\infty$. (This is the usual argument that when left and right inverses both exist, they must be equal).

Theorem 2. Let A be an elliptic $\psi do_\omega : C^\infty(E) \longrightarrow C^\infty(F)$. Then

(i) If f is in some $H^k(E)$ and Af is in $H^s(F)$, then f is in $H^{s+Re(\omega)}(E)$, and there are constants $C_{s,k}$ such that $||f||_{s+Re(\omega)} \leq C_{s,k}(||Af||_s + ||f||_k)$.

(ii) The map $A: H^s(E) \longrightarrow H^{s-Re(\omega)}(F)$ is Fredholm, and there are finitely many sections $g_1,\ldots,g_n$ in $C^\infty(F)$ such that g is in $A(H^s(E))$ iff g is in $H^{s-Re(\omega)}(F)$ and $(g, g_j) = 0$ for $j = 1, \ldots, n$. The g_j are a basis of the null space $N(A^*)$ of A^*.

<u>Proof.</u> Part (i) is contained in Theorem II.6, but it also follows immediately from Lemma 1 by writing

$$f = BAf + (I - BA)f.$$

From this we also get the "a priori" inequality

$$||f||_{s+Re(\omega)} \leq C_{s,k}(||Af||_s + ||f||_k).$$

For part (ii), we observe that I-BA maps $H^s(E)$ continuously into $C^\infty(E)$, hence is a compact operator on $H^s(E)$. The restriction of I-BA to the nullspace N(A) is the identity, and hence N(A) is finite dimensional. To analyze the range of A, we have $AB = I - K$ where K is compact, so $A: H^s(E) \longrightarrow H^{s-Re(\omega)}(F)$ has a closed range which is precisely the orthogonal complement of the null space of $A^*: H^{Re(\omega)-s}(F) \longrightarrow H^{-s}(E)$. But A^* is elliptic, so its null space is spanned by finitely many C^∞ sections $g_1, \ldots, g_n$.

We have proved that A has finite-dimensional null space and a closed range of finite codimension, and this is precisely what it means for A to be Fredholm. Thus Theorem 2 is proved.

If we use $\nu_s(A)$ to denote the nullity of A acting on H^s, i.e., the dimension of the space $\{f \text{ in } H^s(E) : Af = 0\}$, then we have just proved that for any elliptic A, $\nu_s(A)$ and $\nu_s(A^*)$ are finite and independent of s. The difference of these numbers is the index of A,

$$\operatorname{ind}(A) = \nu(A) - \nu(A^*).$$

We shall show that ind(A) depends only on the homotopy class of $\sigma_\omega(A)$. To do so it is convenient to have a topology on the space $\psi do_\omega(E,F)$. We will use the topology of "operators of order $\leq m$"; this topology is defined by the family of norms

$$||A||_{s,s-m} = \sup \frac{||Af||_{s-m}}{||f||_s}.$$

(The interpolation theorem shows that there is an equivalent countable family, e.g. the family $||A||_{n,n-m}$ with n integer.) The space ψdo_ω is not complete in this topology; by working a little harder, and appealing to the theorems on coordinate changes, we could produce a finer topology making ψdo_ω complete, but this simpler operator topology suffices here.

Definition 3. $\Sigma^{\omega}(E,F)$ is the space of C^∞ functions σ on the deleted cotangent bundle $T'(M)$ such that $\sigma(\xi_x)$ is a linear map of E_x into F_x, and $\sigma(t\xi_x) = t^{\omega}\sigma(\xi_x)$ for $t>0$. An element in Σ^{ω} is called a symbol of degree ω.

Theorem 3. The map $A \longrightarrow \sigma(A)$ from $\psi do_\omega(E,F)$ to $\Sigma^{\omega}(E,F)$ has a continuous right inverse, when ψdo_ω is given the topology of operators of order $\leq Re(\omega)$, and $\Sigma^{\omega}(E,F)$ is given the C^∞ topology.

Proof. Cover M with local coordinate patches U_j such that E_{U_j} and F_{U_j} have charts τ_j and τ_j'; we can suppose that the induced diffeomorphisms $\bar{\tau}_j$ and $\bar{\tau}_j'$ of U_j into R^ν are the same. Take a partition of unity $\Sigma\,\varphi_j = 1$ with φ_j in $C_c^\infty(U_j)$, and take ψ_j in $C_c^\infty(U_j)$ such that $\varphi_j\psi_j = \varphi_j$. Then, using τ_j and τ_j', a function σ in $\Sigma^{\omega}(E,F)$ corresponds to a unique matrix $a^j(x,\xi)$, x in $\bar{\tau}_j(U_j)$, a^j homogeneous of degree ω. Setting $\tilde{\varphi}_j(\chi(x)) = \varphi_j(x)$, and taking θ in $C^\infty(R^\nu)$ with $\theta = 1$ for large ξ, we get a matrix $Op(\tilde{\varphi}_j(x)a^j(x,\xi)\theta(\xi))\tilde{\psi}_j(x)$, which yields a well-defined operator A_j in $\psi do_\omega(E,F)$ with $\sigma_\omega(A_j) = \varphi_j\sigma$. Thus $\sigma_\omega(\Sigma A_j) = \sigma$.

We denote by $Op(\sigma)$ this operator ΣA_j. The actual mapping Op is rather arbitrary, depending on the choice of τ_j, τ'_j, φ_j, ψ_j, and θ. The continuity of Op follows on observing that the steps from σ to a^j to $Op(\tilde{\varphi}_j a^j \theta)\tilde{\psi}_j$ to A_j are continuous from Σ^ω to $S^{Re(\omega)}$ to operators of order $\leq Re(\omega)$ on R^ν to operators of order $\leq Re(\omega)$ on $C^\infty(E)$.

Theorem 3 shows that there are many operators in $\psi do_\omega(E,F)$. One consequence of this is:

<u>Theorem 4.</u> Suppose A_0 and A_1 are elliptic, and $\sigma^0 = \sigma_\omega(A_0)$ is homotopic to $\sigma^1 = \sigma_\omega(A_1)$, in the sense that there is a continuous map $t \longrightarrow \sigma^t$ of the interval $[0,1]$ into $\Sigma^\omega(E,F)$ such that $\sigma^t(\xi_x)$ is invertible for all t and all ξ_x. Then $ind(A_0) = ind(A_1)$.

<u>Proof.</u> We define a homotopy of A_0 to A_1 as follows:

$$A_t = (1-3t)A_0 + 3t\ \mathrm{Op}(\sigma^0), \quad 0 \le t \le 1/3$$

$$A_t = \mathrm{Op}(\sigma^{3t-1}), \quad \frac{1}{3} \le t \le 2/3$$

$$A_t = (3-3t)\mathrm{Op}(\sigma^1) + (3t-2)A_1, \quad 2/3 \le t \le 1.$$

Then $\sigma(A_t) = \sigma^0$ for $0 \le t \le 1/3$, $\sigma(A_t) = \sigma^1$ for $2/3 \le t \le 1$, and $\sigma(A_t) = \sigma^{3t-1}$, so A_t is elliptic for every t, and $t \longrightarrow A_t$ is a continuous map into bounded Fredholm operators from H^s to $H^{s-\mathrm{Re}(\omega)}$. It is a basic analytic fact about Fredholm operators that the index of two homotopic operators is the same, so the proof of the Corollary is complete.

We have just shown that the index of two operators of the same degree ω depends only on the homotopy class of the symbols. Actually, this is true even if the two operators have different degrees. For this and other purposes it is convenient to have operators on $C^\infty(E)$ that play the role of the operators Λ^t on Euclidean space.

<u>Lemma 2.</u> Given any real t, there is a self-adjoint Λ_t in $\psi do_t(E)$ such that $\Lambda_t f = 0 \Longrightarrow f = 0$, and $\sigma_t(\Lambda_t)(\xi_x)$ is a positive multiple of the identity in E_x for all $\xi_x \neq 0$. Λ_t is an invertible map of $H^s(E)$ onto $H^{s-t}(E)$ for all s, and Λ_t^{-1} is in $\psi do_{-t}(E)$.

Proof. For $t=0$, take $\Lambda_0 = I$. For $t>0$, choose a symbol σ in $\Sigma^{t/2}(E,E)$ such that $\sigma(\varepsilon_x)$ is a positive multiple of the identity in E_x, for all x, and set $\Lambda_t = Op(\sigma)Op(\sigma)^* + I$. Then Λ_t is self-adjoint, $(\Lambda_t f,f) \geq (f,f)$, and $\sigma_t(\Lambda_t) = \sigma\sigma^* = \sigma^2$, so Λ_t satisfies all the conditions in the first sentence of the Lemma. Further, Λ_t is elliptic, and Λ_t and Λ_t^* have zero null spaces, so by Theorem 2 (ii), Λ_t maps $H^s(E)$ isomorphically onto $H^{s-t}(E)$, and by the closed graph theorem there is a continuous inverse $\Lambda_t^{-1}: H^{s-t}(E) \longrightarrow H^s(E)$. It remains only to prove that Λ_t^{-1} is a ψdo_{-t}. By Lemma 1 there is a ψdo_{-t}, B, such that $S=B\Lambda_t - I$ is in $\psi do_{-\infty}(E)$. In particular, S maps H^s continuously into H^{s+k} for every s and k. Hence $S\Lambda_t^{-1} = B-\Lambda_t^{-1}$ does the same. It follows from Theorem I.3, Corollary, that $B-\Lambda_t^{-1}$ has a C^∞ kernel, i.e. in every local coordinate system it is representable in the form $(B-\Lambda_t^{-1})f(x) = \int k(x,y)f(y)dy$, where $k(x,y)$ maps E_y linearly into E_x, and k is C^∞ in (x,y). Hence, by Theorem III.1, $B-\Lambda_t^{-1}$ is a ψdo_J for every J, with $\sigma_J(B-\Lambda^{-1}) = 0$. We conclude that Λ_t^{-1} is a ψdo_{-t} and $\sigma_{-t}(\Lambda_t^{-1}) = \sigma_{-t}(B) = \sigma_t(\Lambda_t)^{-1}$.

For $t<0$, we can set $\Lambda_t = (\Lambda_{-t})^{-1}$.

Theorem 4 showed that ind(A) depends only on the homotopy class of $\sigma_\omega(A)$. Lemma 2 shows that the degree ω is also irrelevant (at least for ω real), in the following sense. Suppose A_0 is a ψdo_ω and A_1 a ψdo_μ, $\omega \leq \mu$, and $\sigma_\mu(\Lambda_{\mu-\omega} A_0)$ is homotopic to $\sigma_\mu(A_1)$ as in Theorem 4; then $\text{ind}(A_0) = \text{ind}(A_1)$, for $\Lambda_{\mu-\omega}$ is invertible, so $\text{ind}(A_0) = \text{ind}(\Lambda_{\mu-\omega} A_0) = \text{ind}(A_1)$, by Theorem 4.

We conclude this chapter with some further results related to ellipticity, leading to the result that the algebra $\psi do_0(E,E)$ is closed under the application of analytic functions. An important ingredient in this development is the following result, which could be described as Hörmander's version of Gohberg's lemma. (See Hörmander [2] and Gohberg [1].)

Lemma 3. Suppose φ is in $C^\infty(E)$, g is in $C^\infty(M)$, g is real, and dg vanishes nowhere on the support of φ. Then for every A in $\psi do_\omega(E,F)$ we have, as $\lambda \longrightarrow +\infty$,

$$||A(\varphi e^{i\lambda g}) - \sigma_\omega(A)(\cdot\,, \lambda dg\,)\,\varphi e^{i\lambda g}||_0 = O(\lambda^{\text{Re}(\omega)-1}) \tag{1}$$

and

$$||\varphi e^{i\lambda g}||_t = O(\lambda^t) \quad \text{for every real } t. \tag{2}$$

(Note that $\sigma_\omega(A)(\cdot\,, \lambda dg\,)\,\varphi$ denotes the function $x \longrightarrow \sigma_\omega(A)(x, \lambda dg_x)\varphi(x)$.)

Proof. We treat the case where A acts on $C^\infty(M)$. The case of vector bundles differs only in that we work with matrices and local trivializations of E and F.

By a partition of unity, the lemma is reduced to the study of $[A\varphi e^{i\lambda g}](x)$ where x and the support of φ both lie in the domain U of some chart. Since we are free to alter g off the support of φ, we can, by further restricting U, assume that g is the first component of a chart $\chi: U \longrightarrow R^\nu$. Thus we are reduced to the study of $Op(a)\varphi e^{i\lambda g}$ in R^ν, where $g(x) = x_1 = \langle x, \xi \rangle$, $\xi = (1,0,\ldots,0)$, $a \sim \Sigma a_{\omega-j}$, and

$$a_\omega(\chi(y), \lambda\xi) = \sigma_\omega(A)(y, \lambda dg_y).$$

By Lemma II.1,

$$|x^\alpha[e^{-i\lambda g}Op(a)e^{i\lambda g} - a_\omega(x,\lambda\xi)]\varphi(x)| = O(|\lambda\xi|^{Re(\omega)-1}),$$

which yields immediately the estimate (1). For (2), we have $\widehat{\varphi e^{i\lambda g}}(\eta) = \hat{\varphi}(\eta - \lambda\xi)$, and

$$\int |\hat{\varphi}(\eta - \lambda\xi)|^2 (1+|\eta|^2)^t d\eta = \int |\hat{\varphi}(\eta)|^2 (1+|\eta+\lambda\xi|^2)^t d\eta$$

$$\leq C \int |\hat{\varphi}(\eta)|^2)^t (1+|\eta|^2)^t (1+\lambda^2)^t d\eta$$

$$= O(\lambda^{2t}) \,,$$

as was to be proved.

Remark. With a just little more work we could obtain

$$\| e^{-i\lambda g} A(\varphi e^{i\lambda g}) - \sigma_\omega(A)(\cdot \,, \lambda dg)\,\varphi \|_s = O(\lambda^{Re(\omega)-1}) \,,$$

and in fact we could obtain a complete asymptotic expansion of $e^{-i\lambda g} A(\varphi e^{i\lambda g})$, as in Theorem III.3. However, Lemma 3 is adequate for our purposes.

With Lemma 3, we can derive a converse of the a priori inequality in Theorem 2(i).

Theorem 5. Suppose A is in $\#do_\omega(E,F)$, and for some real s and $k<s$ we have the a priori inequality

$$||f||_s \leq C_{s,k}(||Af||_{s-Re(\omega)} + ||f||_k) \text{ , f in } C^\infty(E). \qquad (3)$$

Then $\sigma_\omega(A)(\xi_x)$ is injective for all $\xi_x \neq 0$. (In particular, if dim E = dim F, A is elliptic). In case $Re(\omega) = 0$ and $s = 0$, we have for every x_0 in M, every ξ_0 in T'_{x_0} , and every e_0 in E_{x_0}

$$||e_0|| \leq C_{0,k}||\sigma_\omega(A)(\xi_0)e_0||. \qquad (4)$$

Proof. We begin by taking $Re(\omega)=o=s$ and proving (4). Choose g real with $dg_{x_0} = \xi_0$, and f in $C^\infty(E)$ with $f(x_0) = e_0$, and set $f_n = fe^{i\lambda_n g}$, with $\lambda_n = e^{2\pi n/i\omega}$. Then for φ in $C^\infty(M)$ with support near x_0, we have by Lemma 3

$$||A\varphi f_n - \sigma_\omega(A)(\lambda_n dg)\varphi f_n||_0 \longrightarrow 0, \quad ||\varphi f_n||_k \longrightarrow 0. \qquad (5)$$

If $||\varphi||_0 = 1$ and the support of φ is sufficiently close to x_0, we have

$$||e_o|| \le ||\varphi f_n||_o + \varepsilon$$

and, by the choice of λ_n,

$$||\sigma_\omega(A)(\lambda_n dg)\varphi f_n||_o = ||\sigma_\omega(A)(dg)\varphi f_n||_o$$

$$\le ||\sigma_\omega(A)(\xi_o)e_o|| + \varepsilon.$$

Taking n large and applying (3) and (5), we find

$$||e_o|| \le ||\varphi f_n||_o + \varepsilon \le C_{o,k}(||\sigma_\omega(A)(\xi_o)e_o|| + 3\varepsilon) + \varepsilon.$$

This proves (4), and implies that $\sigma_\omega(A)(\xi_x)$ is injective.

For the case of general ω and s, we use an operator Λ_s on $C^\infty(E)$ and $\Lambda_{s-Re(\omega)}$ on $C^\infty(F)$ as in Lemma 2. Setting $f' = \Lambda_s f$ and $A' = \Lambda_{s-Re(\omega)} A \Lambda_s^{-1}$ reduces the question to the case just treated, and the lemma is proved.

Corollary 1. If A is in $\psi do_{\omega}(E,F)$ and maps $H^{s}(E)$ isomorphically onto $H^{s-Re(\omega)}(F)$ for some s, then A is elliptic, A is an isomorphism of H^{t} onto $H^{t-Re(\omega)}$ for all t, and A^{-1} is in $\psi do_{-\omega}(F,E)$.

Proof. Because of the assumed isomorphism we have (3), hence A is elliptic. Again because of the isomorphism, $\nu(A)$ and $\nu(A^{*})$ vanish, so A maps H^{t} isomorphically onto $H^{t-Re(\omega)}$. It follows that A^{-1} is in $\psi do_{-\omega}(F,E)$, just as in the proof in Lemma 2 that Λ_{t}^{-1} is a ψdo_{-t}.

Corollary 2. If A is in $\psi do_{0}(E,E)$, then the extensions of A to the various spaces $H^{s}(E)$ all have the same spectrum, and this spectrum includes all the eigenvalues of $\sigma_{0}(A)(\xi)$ as ξ ranges over T'(M).

Proof. For any complex λ, $A-\lambda I$ is a ψdo with $\sigma_{0}(A-\lambda I) = \sigma_{0}(A) - \lambda I$. Hence if $A-\lambda I$ is invertible on some H^{s} it is invertible on all H^{s}, and it is elliptic, so λ is not an eigenvalue of $\sigma_{0}(A)$.

Now we come to the result on analytic functions of elements in $\psi do_{0}(E,E)$. An operator in $\psi do_{0}(E,E)$ is, in particular, a bounded operator on the Banach space $H^{0}(E)$. Hence if F is a function analytic on the spectrum of A, we have the bounded operator F(A) on H^{0} defined by the Cauchy integral

$$F(A) = \frac{i}{2\pi} \int_\Gamma F(\lambda)(A-\lambda I)^{-1} d\lambda ,$$

where Γ surrounds the spectrum of A, and F is analytic inside Γ.

Theorem 6. In the situation above, F(A) is a ψdo_0 and $\sigma_0(F(A)) = F(\sigma_0(A))$.

Proof. Let R be the resolvent set of A, $R = \{\lambda : (A-\lambda I)^{-1}$ exists$\}$. By Corollary 2 above, $A-\lambda I$ is elliptic for all λ in R, and the map $\lambda \longrightarrow (A-\lambda I)^{-1}$ is continuous from R into isomorphisms on H^k. We exploit this continuity and the characterization of ψdo's given in Theorem II.7 to show that F(A) is a ψdo_0.

Let τ be a trivialization of E_U, let A_τ be the corresponding transfer of A to a matrix of ψdo's on an open set of R^ν, and let $\sigma(A_\tau) = \Sigma a_{-j}$. Since $A-\lambda I$ is elliptic for λ in the resolvent set R, we can define $b = \Sigma b_{-j}(x,\lambda,\xi)$ for such λ by the formula

$$b \circ [(a_0 - \lambda) + \sum_{j<0} a_{-j}] = 1.$$

In particular, $b_0 = (a_0 - \lambda)^{-1}$. The map $\lambda \longrightarrow b_{-j}(x,\lambda,\xi)$ is continuous from R into C^∞, so we have the homogeneous C^∞ functions

$$c_{-j}(x,\xi) = \frac{i}{2\pi}\int_\Gamma F(\lambda)b_{-j}d\lambda.$$

The theorem will be proved by showing that for every φ, ψ in $C_c^\infty(U)$ such that $\psi = 1$ in a neighborhood of the support of φ, the operator

$$[\varphi F(A)\psi]_\tau - \sum_{j<K} Op(\tilde{\varphi}\theta c_{-j})$$

$$= \frac{i}{2\pi}\int_\Gamma F(\lambda)\{[\varphi(A-\lambda I)^{-1}\psi]_\tau - \sum_{j<K} Op(\tilde{\varphi}b_{-j})\}d\lambda \tag{6}$$

has order $\leq -K$. (Here we have set $\tilde{\varphi}(\tilde{\tau}(x)) = \varphi(x)$, i.e. $\tilde{\varphi}$ is φ transferred to R^ν; and $\theta(\xi)$ is C^∞, vanishing for small ξ and equalling 1 for large ξ). To estimate the order of (6), choose φ_1 and φ_2 in $C_c^\infty(U)$ such that $\varphi\varphi_1 = \varphi$, $\varphi_1\varphi_2 = \varphi_1$, $\varphi_2\psi = \varphi_2$, and set $B(\lambda) = [\tilde{\varphi}_1 \sum_{j<K} Op(\theta b_{-j})\tilde{\varphi}_1]_{\tau^{-1}}$, where the subscript τ^{-1} denotes the transfer from R^ν to M via τ. From Theorem II.1, $\lambda \longrightarrow B(\lambda)$ is continuous from the resolvent set R into bounded operators on H^{t-K}. Further, each of the following maps is continuous from R into bounded operators from H^t to H^{t-K}:

i) $\lambda \longrightarrow \tilde{\varphi} \sum_{j<K} Op(\theta b_{-j})(1-\tilde{\varphi}_1)$

(by Theorem II.1 and Theorem II.2, Corollary, since $\tilde{\varphi}(1-\tilde{\varphi}_1) = 0$)

ii) $\lambda \longrightarrow \tilde{\varphi}\, Op(\sum_{j<K} \theta b_{-j})\tilde{\varphi}_1\, [\varphi_2(A-\lambda I)\psi]_\tau - \tilde{\varphi} I$

(by the same results from Chapter II and the definition of b_{-j})

iii) $\lambda \longrightarrow \varphi B(\lambda)\varphi_2(A-\lambda I)\psi - \varphi I$ (transferring (ii) from R^ν to M)

iv) $\lambda \longrightarrow \varphi_2(A-\lambda I)(1-\psi) = \varphi_2 A(1-\psi)$ (since $\varphi_2(1-\psi) = 0$)

v) $\lambda \longrightarrow \varphi B(\lambda)(A-\lambda I) - \varphi I$

(from (iii) and (iv), noting that $B\varphi_2 = B$ and $\lambda \longrightarrow B(\lambda)$ is continuous)

vi) $\lambda \longrightarrow \varphi B - \varphi(A-\lambda I)^{-1}\psi$

(multiplying through (v) and noting that $B\psi = B$)

vii) $\lambda \longrightarrow \tilde{\varphi}\, \Sigma\, Op(\theta b_{-j})\tilde{\varphi}_1 - [\varphi(A-\lambda I)^{-1}\psi]_T$

(transferring (vi) back to R^ν)

viii) $\lambda \longrightarrow \Sigma\, Op(\tilde{\varphi}\theta b_{-j}) - [\varphi(A-\lambda I)^{-1}\psi]_T$ (from (i) and (vii)).

By the continuity in (viii), (6) is bounded from H^t to H^{t-K}, and Theorem 6 follows from Theorem II.3.

Since ψdo's are continuous operators on C^∞, it follows that if A is a ψdo_0, and F is analytic on the spectrum of A as an operator on L^2, then F(A) is continuous on C^∞. This has an interesting consequence for overdetermined equations. Suppose A is in $\psi do_\omega(E,F)$ with ω real, dim E<dim F, and suppose that $\sigma_\omega(A)(\xi_x)$ is 1-1 for every ξ_x in T'(M).

Theorem 7. In the situation above, there is a projection P in $\psi do_0(F)$ such that $P(H^t(F)) = A(H^{t+s}(E))$ and $P(C^\infty(F)) = A(C^\infty(E))$. Thus P splits every H^t, and C^∞, into the range of A plus a closed complement.

To construct P, first note that $A(H^{t+s}) = A\Lambda_{-s}(H^t)$, so replacing A by $A\Lambda_{-s}$ reduces us to the case s=0; note that $A\Lambda_{-s}(C^\infty) = A(C^\infty)$.

In the next part of the proof we consider A as an operator on H^0, and apply Hilbert space theory. Form A^*A; because $\sigma_0(A)$ has maximal rank, $\sigma_0(A^*A)$ is non-singular, so A^*A is elliptic and has closed range. It follows that A has closed range, and range (A) = range (AA^*). Let P be the orthogonal projection on this range. Since AA^* is self-adjoint and has closed range, zero must be an isolated point in spectrum (AA^*). Hence, for some small $\varepsilon > 0$, AA^* has no spectrum in $0 < |\lambda| < \varepsilon$, and the function

$$F(\lambda) = 0,\ |\lambda| < \varepsilon,\ F(\lambda) = 1,\ |\lambda| > \varepsilon$$

is analytic in a neighborhood of spectrum (AA^*). Hence we can represent our projection P as

$$P = F(AA^*) = \frac{i}{2\pi} \int_{\Gamma} F(\lambda)(AA^* - \lambda I)^{-1} d\lambda, \tag{7}$$

Let $P' = I-P$. Then note from (7) and the choice of ε that

$$PP' = P'P = 0, \; PAA^* = AA^*. \qquad (8)$$

Now we have an orthogonal projection P onto $A(H^0)$, and P is a ψdo by (7) and Theorem 6. It remains only to show that $P(H^t) = A(H^t)$, and $P(C^\infty) = A(C^\infty)$. For this, note that $AA^* + P'$ is an isomorphism on H^0. Hence by Corollary 1 above, A^*A+P' is elliptic, and is an isomorphism on every H^k: given any ψ in H^k, there is a unique φ in H^k such that

$$AA^*\varphi + P'\varphi = \psi, \qquad (9)$$

If $\psi \in P(H^k)$ then $P\psi = \psi$, and multiplying (9) by P and noting (8), we find $AA^*\varphi = \psi$, so $\psi \in A(H^k)$. Conversely, suppose $\psi \in A(H^k)$, and consider two cases:

(i) $k \geq 0$. Then $\psi \in A(H^0) = AA^*(H^0)$, so $P'\psi = 0$.

(ii) $k < 0$. Then $\psi = A\varphi$, and there is a sequence φ_m in H^0 such that $\varphi_m \longrightarrow \varphi$ in H^k. By case (i), $P'A\varphi_m = 0$, so on taking limits $P'\psi = P'A\varphi = 0$.

Thus we have proved that $P(H^k) = A(H^k)$. Since this is true for every k, and $C^\infty = \cup H^k$, it is true for C^∞ as well, and all the claims for P are established.

V. Complex powers of an elliptic operator.

This chapter sketches some results on the complex powers of an elliptic operator A in the simplest general case, i.e., where A is a differential operator acting on a compact manifold without boundary; this case has been treated by T. Burak [1], D. Fujiwara [2], and P. Greiner [1]. The corresponding results when A is a ψdo are given in Seeley [2; see also corrections to that article given in [3]], and, when A is a differential operator with boundary conditions, in [3,4]. These last results have also been announced by P. Greiner [2]. The original results in this direction appear to be due to H. Weyl [1] and T. Carleman [1]; Carleman treats the so-called "zeta function" $\zeta(x,z) = K_z(x,x)$, where K_z is the kernel in the integral representation

$$A^z f(x) = \int K_z(x,y)f(y)dy,$$

which is valid for Re(z) sufficiently negative, and bases his results on a study of the resolvent $(A-\lambda)^{-1}$.

The resolvent enters in a very natural way in view of the Cauchy integral

$$A^z = \frac{i}{2\pi} \int_\Gamma \lambda^z (A-\lambda)^{-1} d\lambda, \tag{1}$$

where Γ is an appropriately chosen contour. We begin by deriving an asymptotic expansion of $(A-\lambda)^{-1}$ for large λ, then study A^z by using the formula (1).

The main idea in studying the resolvent for large values of λ is due to Agmon [1]; the parameter λ enters essentially as a new Fourier transform variable, weighted according to the order of the operator A. If $A = \sum_{|\alpha| \leq \omega} a_\alpha(x) D^\alpha$, we define

$$a_\omega(x,\xi,\lambda) = \sum_{|\alpha|=\omega} a_\alpha(x)\xi^\alpha - \lambda,\ a_j = \sum_{|\alpha|=j} a_\alpha(x)\xi^\alpha \quad \text{for } 0 \leq j < \omega, \tag{2}$$

and have $\sigma(A-\lambda) = \sum_0^\omega a_{\omega-j}$. Thus the parameter λ enters into the homogeneity with weight ω:

$$a_j(x,t,\xi,t^\omega\lambda) = t^j a_j(x,\xi,\lambda). \tag{3}$$

Once the symbol has been set up like this, we approximate the resolvent by following the construction of a parametrix. Define functions $b_{-\omega-j}$ by the asymptotic formula

$$(\Sigma\, b_{-\omega-j}) \circ (\Sigma\, a_{\omega-k}) \sim 1,$$

i.e.

$$\left.\begin{aligned} & b_{-\omega}\, a_{\omega} = 1 \\ & b_{-\omega-j}\, a_{\omega} + \sum_{\ell<j} b^{(\alpha)}_{-\omega-j}\, D^{\alpha}_{x}\, a_{\omega-\ell}/\alpha! = 0, \quad j \geq 0 \end{aligned}\right\} \quad (4)$$

where the sum is for $k + \ell + |\alpha| = j$. The first equation in (4) can be solved when $a_{\omega}(x,\xi,\lambda)$ is non-singular. If this condition holds for some λ and all $\xi \neq 0$, then it holds (by homogeneity) along the half line $\{t\lambda : t > 0\}$. Thus we are led to Agmon's condition (for a ray of minimal growth): Let A be an elliptic differential operator of order ω, $\sigma(A) = \Sigma\, \sigma_{\omega-j}(A)$ its characteristic polynomial, and $\rho_{\theta} = \{\lambda : \arg \lambda = \theta\}$ be a ray in the complex plane. Then ρ_{θ} is a ray of minimal growth if the matrix $\sigma_{\omega}(A)(x,\xi)$ has no eigenvalues on ρ_{θ} for any x and ξ. In view of the interpretation of $\sigma_{\omega}(A)$ as a function on $T'(M)$, this definition makes sense for an operator on a compact manifold M, or, more generally, for any A in $\Psi do_{\omega}(E,E)$, where E is a vector bundle on M and ω is a non-zero real number.

<u>Theorem 1.</u> Suppose that the negative real axis ρ_π is a ray of minimal growth for an elliptic differential operator A of order ω. Then there is a sector S containing ρ_π such that $A-\lambda I$ is invertible for λ sufficiently large in S, and $||(A-\lambda I)^{-1} f||_s = O(||f||_s/|\lambda|)$. Further, if the $b_{-\omega-j}$ are defined by (4) in local bundle coordinates $\tau: E_U \longrightarrow V \times C^p$, and if φ and ψ are in $C_c^\infty(V)$, then for all $J = 0,1,\ldots$,

$$\varphi[\sum_{j<J} Op(b_{-\omega-j}) - (A-\lambda I)_\tau^{-1}]\psi$$

is bounded on H^s with norm $O(|\lambda|^{-1-J/\omega})$, or from H^s to $H^{s+J+\omega}$ with norm $O(1)$, and generally from H^s to $H^{s+\theta(J+\omega)}$ with norm $O(|\lambda|^{(\theta-1)(1+J/\omega)})$, $0 \leq \theta \leq 1$.

In particular, $(A-\lambda I)^{-1}$ has norm $O(1/|\lambda|)$ on H^s. It is easy to see that $||(A-\lambda I)^{-1}||$ cannot decay any faster than $1/|\lambda|$ as $\lambda \longrightarrow \infty$, hence the term "ray of minimal growth"; the resolvent grows as slowly as possible on such a ray.

To prove Theorem 1, note first by compactness that there is a sector S containing ρ_π, and consisting entirely of rays of minimal growth. Henceforth we shall tacitly assume that λ is in this sector S. In view of (2) and (4), $b_{-\omega-j}$ is C^∞ for $(\xi,\lambda) \neq 0$, and homogeneous of degree $-\omega-j$, that is

$b_{-\omega-j}(x,t\xi,t^{\omega}\lambda) = t^{-\omega-j}\, b_{-\omega-j}$. (Notice that $(\xi,\lambda) \neq 0$ does not rule out $\xi = 0$, so it is essential for the differentiability of $b_{-\omega-j}$ that the $a_{\omega-k}$ be C^{∞} at $\xi = 0$, i.e. that the $a_{\omega-k}$ be polynomials in ξ. This is where the assumption that A is a differential operator simplifies the argument). Following the argument of Theorem II.2, we get for φ and ψ in $C_c^{\infty}(V)$ with $\varphi\psi = \varphi$,

$$\mathrm{Op}(\varphi b_{-\omega-j})\ \mathrm{Op}(\psi a_{\omega-k}) =$$

$$\sum_{|\alpha|<r} \mathrm{Op}(\varphi b^{(\alpha)}_{-\omega-j}\, D^{\alpha}_{x}(\psi\, a_{\omega-k})/\alpha!) + \mathrm{Op}(R_r) \ , \qquad (5)$$

where R_r involves the integral form of the Taylor remainder,

$$R_r(x,\eta,\lambda) = \varphi(x) \iint_0^1 \sum_{|\alpha|=r} \frac{r}{\alpha!} \int_0^1 (1-t)^{r-1} b^{(\alpha)}_{-\omega-j}\ (x,\eta+t\sigma,\lambda)dt \cdot$$

$$\cdot D^{\alpha}_{x}\ \psi a_{\omega-k}(\sigma,\eta,\lambda)d\sigma \ . \qquad (6)$$

From the homogeneity and differentiability of the $a_{\omega-k}$ and $b_{-\omega-j}$, we get for the factors in (6)

$$|D_x^\beta\, b_{-\omega-j}^{(\alpha)}| \le C_{\alpha\beta}(|\eta + t(\xi-\eta)| + |\lambda|^{1/\omega})^{-\omega-j-|\alpha|}$$

$$\le C_{\alpha\beta}(|\eta| + |\lambda|^{1/\omega})^{-\omega-j-|\alpha|}(1+|\xi-\eta|\,|\lambda|^{-1/\omega})^{\omega+j+|\alpha|} ,$$

$$|D_x^\alpha\, \psi a_{\omega-k}| \le C_K(1+|\xi-\eta|)^{-K}(|\eta| + |\lambda|^{1/\omega})^{\omega-k}$$

Hence, when $|\lambda| \ge \varepsilon > 0$, we get for $r = J-j-k$

$$|D_x^\beta\, R_r(x,\eta,\lambda)| \le C(|\eta| + |\lambda|^{1/\omega})^{-J}$$

$$\le C(\varepsilon + |\eta|)^{-\theta J}\ |\lambda|^{(\theta-1)J/\omega} , \quad \text{for all } 0 \le \theta \le 1.$$

so $||Op(R_r)||_{s,s+\theta J} = O(|\lambda|^{(\theta-1)J/\omega})$, $0 \le \theta \le 1$, $\lambda \longrightarrow \infty$.

Returning to (5), taking $r = J-j-k$, summing over $0 \le j < J$ and $0 \le k \le \omega$, noting (4) and recalling that $\varphi D^\alpha \psi a = \varphi D^\alpha a$ (since $\omega\psi = \varphi$), we get

$$\left\| \sum_{j<J} Op(\varphi b_{-\omega-j}) \sum Op(\psi a_{\omega-k}) - M_{\varphi} \right\|_{s,s+\theta J} = O(|\lambda|^{(\theta-1)J/\omega}),$$

$$0 \leq \theta \leq 1.$$

Thus $\varphi \sum OP(b_{-\omega-j}) \psi$ is a local parametrix for $A-\lambda$. Taking a partition of unity $\sum \varphi_{\ell} = 1$ and transferring such local parametrixes to M, we obtain a global parametrix B such that

$$\|B(A-\lambda) - I\|_{s,s+\theta J} = O(|\lambda|^{(\theta-1)J/\omega}). \tag{7}$$

By the usual appeal to the geometric series, it follows that $A-\lambda$ has a left inverse for λ large in S, and repeating this construction for $A^{*} - \bar{\lambda}$ we get a right inverse for $A-\lambda$. Moreover, from the estimates on $b_{-\omega-j}$,

$\|B\|_{s,s+\theta\omega} = O(|\lambda|^{\theta-1})$, $0 \leq \theta \leq 1$, and it follows easily that this same estimate holds with B replaced by $(A-\lambda)^{-1}$. Finally, multiplying (7) by $(A-\lambda)^{-1}$, we get

$$\|B-I\|_{s,s+\theta(J+\omega)} = O(|\lambda|^{(\theta-1)(1+J/\omega)}).$$

Thus, because of the way that B is constructed, we get precisely the result stated in Theorem 1.

Now we return to the study of

$$A^z = \frac{i}{2\pi} \int_\Gamma \lambda^z (A-\lambda I)^{-1} d\lambda. \qquad (1)$$

We assume that A is invertible, and has no spectrum on the negative real axis. We take Γ running from $\lambda = -\infty$ to $\lambda = -\varepsilon$, then clockwise around the circle $\{\lambda: |\lambda| = \varepsilon\}$, then from $-\varepsilon$ to $-\infty$ again; ε is taken so small that Γ "surrounds" the spectrum of A. We have shown that $||(A-\lambda I)^{-1}||_{s,s} = O(1/|\lambda|)$, so (1) converges for $\mathrm{Re}(z) < 0$ to a bounded operator on H^s, and we can show directly from the contour integral that $A^{z+\omega} = A^z A^\omega$, and $A^{-1}A = I$. To go further, we want to replace $(A-\lambda I)^{-1}$ in (1) by Σ Op $(b_{-\omega-j})$, but there is a minor difficulty along $|\lambda| = \varepsilon$, where λ passes out of the sector S in which $a_\omega(x,\xi,\lambda)$ is invertible. To evade this, we appeal to the ellipticity of A to find a $\delta > 0$ such that $\sigma_\omega(A)(x,\xi)$ has no eigenvalue of absolute value $\leq \delta|\xi|$. Hence if we take a C^∞ function $\theta(\xi,\lambda)$, equalling 1 for $|\xi|^2 + |\lambda|^{2/\omega} \geq 1$ and 0 for $|\xi|^2 + |\lambda|^{2/\omega} \leq 1/2$; and if we take the ε defining the path Γ sufficiently small, then $\theta\, b_{-\omega-j}$ is well-defined and C^∞ for all (ξ,λ) with λ on Γ. Let

$$c_j(x,\xi,z) = \frac{i}{2\pi}\int_\Gamma \lambda^z \theta b_{-\omega-j}(x,\xi,\lambda)d\lambda \ . \qquad (8)$$

It is easy to check that $c_j(x,t\xi,z) = t^{\omega z-j}\, c_j(x,\xi)$ for $t \geq 1$, $|\xi| \geq 1$. Further, applying Theorem 1 with $\theta = J/(J+\omega)$, we find that $\varphi A^z\psi - \varphi \sum_{j<J} Op(c_j)\psi$ is bounded from H^s to H^{s+J}, for $Re(z) < 0$. It follows from Theorem II.9 that A^z is a $\psi do_{\omega z}$ for $Re(z)<0$, and in local coordinates $\sigma(A^z) = \Sigma c_j$ $|\xi| \geq 1)$. The formula $AA^z = A^{z+1}$ $(Re(z+1)<0)$ allows us to extend A^z analytically to all z, and $\sigma(A^z)$ likewise.

When $Re(\omega z) < -\nu$, A^z has a continuous kernel, i.e. in local coordinates $A^z f(x) = \int K_z(x,y)f(y)dy$, where $K_z(x,y)$ is continuous. (See Theorem I.3.) We can prove: that $K_z(x,y)$ is analytic in z for $Re(\omega z) < -\nu$; that for each fixed x and y, $K_z(x,y)$ extends to a meromorphic function in the complex z-plane (in fact, when $x \neq y$, the extension is an entire function); that all the poles of $K_z(x,x)$ are simple and have explicitly computable residues; and finally that the continuations $K_z(x,x)$ can be explicitly computed for $z = 0,1,2,\dots$. (If A were not a differential operator, this last claim would hold only for $z = 0$.)

The following question arises: what is the connection between this analytic continuation of K_z and the kernel of A^z? The answer is: <u>off</u> the diagonal, $K_z(x,y)$ is precisely the kernel of A^z, whereas <u>on</u> the diagonal $K_z(x,x)$ sheds no particular light on the kernel of A^z. For instance, when $z=0$, $A^z = I$, so $K_0(x,y) = 0$, if $x \neq y$; however, we find a value for $K_0(x,x)$ which depends on A, whereas $A^0 = I$ does not depend on A.

For the details of this analysis we refer to Seeley [2] (and the corrections in [3]). However, the main points can be easily explained.

First of all, if we set $C_z = \frac{i}{2\pi} \int_\Gamma \lambda^z B\, d\lambda$, where B is the parametrix for $(A-\lambda)^{-1}$, we find by taking $\theta = (\nu+1)/(J+\omega)$ in Theorem 1 that $A^z - C_z$ extends analytically to $\mathrm{Re}(z) < \frac{J-\nu-1}{\omega}$ as a bounded operator from H^s to $H^{s+\nu+1}$. Hence, if L_z is the kernel of C_z, the kernel $K_z - L_z$ of $A^z - C_z$ extends analytically to $\mathrm{Re}(z) < \frac{J-\nu-1}{\omega}$, the extension being continuous in (x,y,z). Hence, any poles or discontinuities of K_z are the same as the poles and discontinuities of L_z.

Second, the essential contributions to the kernel $L_z(x,y)$ of C_z are the integrals

$$(2\pi)^{-\nu} \int_{|\xi|\geq 1} e^{i\langle x-y,\xi\rangle} c_j(x,\xi,z)d\xi \ .$$

For $x \neq y$, it is easy to continue these as entire functions in z, simply by an integration by parts. For x=y, the integrals can be evaluated in polar coordinates, yielding

$$\frac{1}{j-\nu-\omega z} \int_{|\xi'|=1} c_j(x,\xi',z)d\xi' . \tag{9}$$

This shows clearly a simple pole at $z = \frac{j-\nu}{\omega}$

Third, when $z=0,1,2,\ldots$, then A^z is a differential operator, hence for $x \neq y$ the kernel $K_z(x,y)$ vanishes for these values of z. Thus, $K_z(x,x)$ is precisely the discontinuity of $K_z(x,y)$ at y=x. We can compute the discontinuity of L_z for these values of z, hence by the first of these three remarks we obtain the actual value of $K_z(x,x)$ for $z=0,1,2,\ldots$.

Precisely,

$$K_z(x,x) = \frac{(-1)^z}{\omega(2\pi)^\nu} \int_{|\xi'|=1} \int_0^\infty t^z b_{-\omega-z\omega-\nu}(x,\xi',-t)\,dt\,d\xi', \quad z=0,1,2,\dots \tag{10}$$

The main application of these results comes when A is a positive semi-definite operator. Then there are eigenfunctions $A\varphi_j = \lambda_j\varphi_j$, and we have

$$\xi_A(-z) = \text{trace}\,(A^z) = \Sigma\lambda_j^z = \int_M \text{trace}\, K_z(x,x)dx, \ \text{Re}(\omega z) < -\nu .$$

Hence we can obtain the residues and certain values of $\Sigma\lambda_j^z$ by taking the trace of the forms in (9) and (10) , and integrating over M.

In particular, suppose D is any elliptic differential operator, and let N_λ be the eigenspace of D^*D with eigenvalue λ, and M_λ be the eignespace of DD^* with eigenvalue λ. Then for $\lambda\neq 0$, D maps N_λ isomorphically into M_λ and D^* maps M_λ isomorphically into N_λ, so the non-zero eigenvalues of D^*D and DD^* are precisely the same, counting multiplicity. Hence $\xi_{I+D^*D} - \xi_{I+DD^*}$ is constant and equals the index of D.

Since both terms can be evaluated at $z=0$, we obtain a formula for the index. However, the formula is very hard to compute except when ν is very small; as (10) shows, it involves $b_{-\omega-\nu}$, and $b_{-\omega-\nu}$ is obtained by ν recursions involving $a_{\omega}, a_{\omega-1}, \ldots,$ and their derivatives of order $\leq \nu$. In spite of these complications, one result comes out easily: when ν is odd and D is an elliptic differential operator, then index(D) = 0. For it is easy to check in this case that the $b_{-\omega-\nu}$ computed for either $A = I+D^*D$ or $A = I+DD^*$ is an odd function of ξ, so the integral (10) vanishes.

Three final remarks. First, most of the applications of pseudo-differential operators do not exploit in an essential way the expansion in homogeneous functions that is characteristic of "classical" pseudo-differential operators; in fact, Hormander [3] gives up the homogeneity to provide a framework for general results on hypoelliptic operators. For the computation of values and residues of $K_z(x,x)$ by formulas (10) and (9), however, the homogeneity does play an essential role.

Second, there are analogous results on $(A_B)^z$ when A_B is a realization of an elliptic differential operator with domain defined by homogeneous boundary conditions $Bu = 0$. (Seeley [3],[4] and Grenier [2].) It would be interesting to fit $(A_B)^z$ for non-integer z into the framework provided by the work of Boutet de Monvel, as outlined at this conference.

Third, in case A is the Laplacian Δ on a Riemannian manifold, the symbol $\sigma(\Delta-\lambda) = a_2 + a_1 + a_0$ is derived from the metric, so the same holds for the terms b_{-2-j} of the parametrix, and likewise for the forms giving the residues and values of $\zeta_\Delta(z)$. But, just as in the case of the index, the formulas for these forms are inordinately complicated. Berger [1] and McKean and Singer [1] have found simpler expressions for some of them, and it would be interesting to see if further progress can be made in this direction.

VI. Boundary value problems.

The most familiar boundary value problem is the Dirichlet problem: Given a domain G in R^2, a function f in G, and a function g on the boundary ∂G, find u such that $\Delta u = f$ in G, and $u = g$ on ∂G. Classically, the solution is found in two stages. First extend f in any convenient way to R^2, and take the Newtonian potential $v(x) = c \int P(x-y)f(y)dy$, where $P(x) = c \log|x|$. Since $\Delta v = f$, the substitution $u = v+w$ leads to

$$\Delta w = 0 \text{ in } G, \qquad w = g_1 \text{ on } \partial G \qquad (\text{where } g_1 = g-v).$$

This problem is converted into an integral equation on ∂G by the classical method of "double layer potentials", by writing w in the form

$$w(x) = \int_{\partial G} \partial P(x-y)/\partial n_y \, f(y) dS_y ,$$

where $\partial/\partial n_y$ is the normal derivative with respect to y, and dS_y is the area element on ∂G. For the boundary values of w one finds

$$w(x) = cf(x) + \int_{\partial G} K(x,y)f(y)dy, \quad x \text{ in } \partial G ,$$

where K is a kernel with an integrable singularity (actually a pseudo-differential kernel of degree $2_{-\nu}$). Setting $w(x) = g_1(x)$ leads to an integral equation for f that is solved by the Fredholm theory. (See Riesz and Nagy [1,No. 81].)

We will state two theorems that make it possible to repeat this procedure for general elliptic operators A, and show some of the applications of the results.

Suppose that M^+ is a compact C^∞ manifold with boundary $X = \partial M$; we take M^+ to be an open submanifold of its "double" M, with X a compact hypersurface in M. Given a bundle E over M^+, we can extend E over M, and near X we can represent it as $(-1,1) \times E_X$, so that sections u of E can be written $u(x,t)$, with x in X, t in $(-1,1)$, $u(x,t)$ in E_X. When such a representation is chosen we can define, for $-1 < \varepsilon < 1$, restriction maps $R_\varepsilon^\omega: C^\infty(E) \longrightarrow C^\infty(E_X)^\omega$ by

$$[R_\varepsilon^\omega u](x) = [u(x,\varepsilon),\dots,D_t^{\omega-1}u(x,\varepsilon)], \; D_t = -i\, \partial/\partial t .$$

It is well known that the map $(\varepsilon,u) \longrightarrow R^{\omega}_{\varepsilon}u$ is separately continuous from $(-1,1) \times H^{s}(E)$ into $\sum_{0}^{\omega-1} H^{s-j-1/2}(E_X)$, for $s > \omega - 1/2$.

We introduce the following spaces:

$H^{s}(E_{M^+})$ = the restrictions to M^+ of elements in $H^{s}(E)$

$H^{s}_{0}(E_{M^+})$ = the elements in $H^{s}(E)$ with support in M^+

$N(A,s) = \{u \text{ in } H^{s}(E_{M^+}) : Au = 0 \text{ in } M^+\}$

$N_0(A,s) = \{u \text{ in } H^{s}_{0}(E_{M^+}) : Au = 0\}$

$R_0(A,s) = \{f \text{ in } \Sigma\, H^{s-j-1/2}(E_X) : f = \lim_{\varepsilon \to 0} R^{\omega}_{\varepsilon}u$ for

some u in $N(A,s)\}$.

<u>Theorem 1.</u> Let A be an elliptic differential operator on $E_{M^+ \cup X}$. Then

(i) $N_0(A,s)$ is a finite-dimensional subspace of $C^\infty(E)$; in particular, it is independent of s, and will be denoted $N_0(A)$.

(ii) There is a map P from $C^\infty(E_X)^\omega$ into $C^\infty(E_{M^+})$ which, for each s, extends to a continuous map of $\Sigma H^{s-j-1/2}(E_X)$ into N(A,s).

(iii) N(A,s) is the direct sum of $P(\Sigma H^{s-j-1/2})$ and $N_0(A)$.

(iv) For any f in $\Sigma H^{s-j-1/2}(E_X)$, $\lim_{\varepsilon \to 0+} R_\varepsilon^\omega P f$ exists in norm; call this limit $P^+ f$.

(v) P^+ is a ψdo, and is a projection onto $R_0(A,s)$.

This result is proved in Seeley [5]. Similar results are obtained by Hörmander [4] and Boutet de Monvel [1], the main difference being that here we make the operators P and P^+ do the best possible job, whereas the other works achieve this only on the level of symbols. This is more a question of convenience than an essential difference, however. The proof in Seeley [5] contains an error which is corrected in the appendix below.

The projection $P^+ : \Sigma H^{s-j-1/2} \longrightarrow \Sigma H^{s-j-1/2}$ has a natural representation as an $\omega \times \omega$ matrix (P^+_{jk}), where P^+_{jk} is a ψdo_{j-k}. Denote by p^+_o the matrix $(\sigma_{j-k}(P^+_{jk}))$ of top symbols. This top symbol p^+_o is also a projection; in fact, if we write A along the boundary X in the form $A = \Sigma A_j D_t^{\omega-j}$, where A_j is a differential operator on E_X of order $\leq j$, then $p^+_o(\xi_x)$ is a projection onto the initial data of the solutions of

$$\Sigma \sigma_j(A_j)(\xi_x) D_t^{\omega-j} u = 0, \quad u(+\infty) = 0.$$

The complementary projection $I - p^+_o(\xi_x)$ projects onto the corresponding solution space with $+\infty$ replaced by $-\infty$.

Theorem 1 covers the "surface potentials" arising in elliptic boundary problems. For the "volume potentials" we have the following result whose proof is sketched in the appendix below.

<u>Theorem 2.</u> Let A be an elliptic differential operator:

$C^\infty(E_{M^+UX}) \longrightarrow C^\infty(F_{M^+UX})$. Then there is a $\psi do_{-\omega}$ C such that

i) $C^*R_0^{\omega *} \mid M^+$ is continuous from $\Sigma H^{-s+j+1/2}(F_X)$ to $H^{\omega-s}(E_{M^+})$ for for each s, and the same holds with M^+ replaced by M^-;

(ii) if u is in $H^s(E_{M^+})$ for some $|s|\leq k$, and orthogonal to $N_0(A^*)$, then $ACE_k u = u$ in M^+.

Here $R_0^{\omega *}: \sum_0^{\omega-1} H^{-s+j+1/2}(E_X) \longrightarrow H^{-s}(E_{M^+})$ is the adjoint of R_0^ω, and E_k is an extension map: $H^s(E_{M^+}) \longrightarrow H^s(E)$ for $|s|\leq k$, given by (1) below. Since C is a $\psi do_{-\omega}$, the map $C^*R_0^{\omega *}$ is automatically continuous from $\Sigma H^{-s+j+1/2}$ to $H^{\omega-s}$ for $s<1/2$. The point of Theorem 2(i) is that for f in $\Sigma H^{-s+j+1/2}$, $C^*R_0^{\omega *}f \mid M^+$ ends up in $H^{\omega-s}(E_{M^+})$ even if $s\geq 1/2$, and the map so obtained is continuous.

In part (ii) the extension map E_k is obtained by "reflections",

$$E_k u(x,t) = \begin{cases} u(x,t), & \text{if } t \geq 0 \\ \varphi(t)\, \Sigma_1^{2k} a_j u(x,-jt) & \text{if } t < 0 , \end{cases} \tag{1}$$

where $\Sigma_1^{2k} (-j)^n a_j = 1$ for $n = -k$, $1-k, \ldots, k-1$, and φ is C^∞, $\varphi(t) = 1$ for $t \geq 0$, $\varphi(t) = 0$ for $t \leq -1$.

Now suppose we are given a ψdo, call it B, mapping $\sum_0^{\omega-1} H^{s-j-1/2}(E_X)$ into sections of some bundle G over X. B has a natural representation as an ω-tuple $\{B_0, \ldots, B_{\omega-1}\}$ of operators from $C^\infty(E_X)$ to $C^\infty(G)$, and we assume that B_j has order $\omega-j-1$. For a differential operator $A: C^\infty(E_{M^+}) \longrightarrow C^\infty(F_{M^+})$, we denote by (A,B) the map $u \longrightarrow (Au, BR_0^\omega u)$ from $H^s(E_{M^+})$ to $H^{s-\omega}(F_{M^+}) + H^{s-\omega+1/2}(G)$, $s > \omega - 1/2$. (The usual ways of forming boundary value problems can always be cast in this form; for example, to assign Dirichlet data for an operator A of order 2, take $B_0 = \Lambda_1$ and $B_1 = 0$, where Λ_1 is an invertible ψdo$_1$ such as we constructed in Chapter IV.)

<u>Definition 1.</u> The boundary problem (A,B) is <u>elliptic</u> if and only if A is elliptic and, for each $\xi_x \neq 0$, $b_o(\xi_x)$ is an isomorphism of the range of $p_o^+(\xi_x)$ onto the fibre G_x.

In view of the characterization of p_o^+ given above, it is easy to show that this definition is equivalent to the more familiar versions of the Lopatinsky condition.

For elliptic boundary problems we construct a parametrix with the help of

<u>Lemma 1.</u> If (A,B) is an elliptic boundary system, then there exists a ψdo, call it D, mapping $H^{s-\omega}(G)$ into $\Sigma H^{s-j}(E_X)$, such that $DBP^+ - P^+$ and $BP^+D - I$ have degree $-\infty$.

<u>Proof.</u> Just as in the construction of a parametrix in Chapter IV, we work in local coordinates; then it is trivial to transfer the result to X by a partition of unity. In some local coordinates, let $p^+ = \sigma(P^+)$ and $e = \sigma(BP^+)$; these are complete symbols, not just leading terms. Then we have to find a solution d of the equations

$$e \circ d = I \qquad (2)$$

$$d \circ e = p^+. \qquad (3)$$

Beginning with (3), note that by the ellipticity assumptions there is a d_o such that $d_o e_o = p_o^+$. Then

$$d_o \circ e = p^+ - r , \tag{4}$$

where $r = (r_{jk})$ is a matrix in which r_{jk} involves terms of degree $\leq j-k-1$. Hence the jk entry in $r \circ r$ has degree $\leq j-k-2$, etc., so the geometric series $I + r + r \circ r + \ldots$ defines a symbol. Since $e = \sigma(BP^+) = \sigma(B) \circ p^+$, and p^+ is a projection, we have

$$e \circ p^+ = e, \tag{5}$$

hence from (4)

$$r \circ p^+ = r. \tag{6}$$

In view of (4) and (6), $d = (I + r + r \circ r + \ldots) \circ d_o$ solves (3).

Similarly, starting with a d_o such that $e_o d_o = I$, we obtain a d' satisfying $e \circ d' = I$. It follows from a multiplication on the left by d that $p^+ \circ d' = d$, hence (2) follows from (5).

To finish the proof, take a partition of unity $\Sigma\, \varphi_j = 1$ and "covering functions" ψ_j in $C_c^\infty(U_j)$ such that $\psi_j \varphi_j$, and such that E_X and G are trivial when restricted to U_j. Then if D_j has, in local coordinates in U_j, the symbol just constructed, it follows that $\sigma(\varphi_j D_j \psi_j)BP^+ = BP\varphi_j P^+)$; hence if $D = \Sigma\, \varphi_j D_j \psi_j$, $DBP^+ - P^+$ has degree $-\infty$. Similarly, if $D' = \Sigma\, \psi_j D_j \varphi_j$, then $BP^+D' - I$ has degree $-\infty$. If follows that $P^+D' - D$ has degree $-\infty$, hence

$$BP^+D - I = BP^+D' - I + BP^+(D - P^+D')$$

has degree $-\infty$, and Lemma 1 is proved.

Corollary 1. For each real s and t there is a constant C such that

$$||P^+||_s \leq C(||BP^+f||_{s-\omega+1} + ||f||_t) ,$$

where $||f||_t$ is the norm of f in $\Sigma\, H^{t-j}(E_X)$, and $||BP^+f||_{s-\omega+1}$ the norm in $H^{s-\omega+1}(G)$.

Proof. $P^+f = DBP^+f - Sf$, where S has degree $-\infty$ and D is continuous from $H^{s-\omega+1}(G)$ to $\Sigma\, H^{s-j}(E_X)$.

Corollary 2. When B is restricted to the range of P^+, its null space is finite dimensional and C^∞, and its range is the orthogonal complement of finitely many C^∞ sections of G.

Proof. The null space of $DBP^+ = p^+ - S$ contains the null space of B restricted to the range of P^+. The range of BP^+ includes the range of $BP^+D = I - S'$. Since S and S' have degree $-\infty$, corollary 2 follows.

Now we obtain the standard result on elliptic boundary problems.

Theorem 3. Let (A,B) be an elliptic boundary problem. Then for $s \geq \omega$, (A,B) maps $H^s(E_{M^+})$ into $H^{s-\omega}(E_{M^+}) \oplus C^\infty(G)$.

If B involves only the Cauchy data of order < m, then the same results hold for $s \geq m$.

Proof. We have already noted the continuity of (A,B). By Theorem 1, the null space of (A,B) is the direct sum of $N_0(A)$ and the space

$$\{Pf: f = P^+f,\ Bf = 0,\ f \text{ in } \Sigma\, H^{s-j-1/2}\}$$

so the claim about the null space of (A,B) follows from Corollary 2.

Considering now the range of (A,B), note that the first condition that (u,g) be in the range of (A,B) is that u be orthogonal to $N_0(A^*)$, where A^* is the formal adjoint of A. Supposing this is satisfied, let $v = CE_k u$ as in Theorem 2; then $Av = u$, and we are reduced to solving

$$Aw = 0, \quad BR_0^{\omega} w = g - BR_0^{\omega} v . \tag{7}$$

By Theorem 1, this can be solved if and only if $g - BR_0^{\omega} v$ is in the range of BP^+; for if $g - BR_0^{\omega} v = BP^+ f$, then $w = Pf$ solves (7), while conversely if w solves (7), then $R_0^{\omega} w = P^+ R_0^{\omega} w$, so $g - BR_0^{\omega} v = BP^+ R_0^{\omega} w$. Thus, by Corollary 2 above, (7) has a solution if and only if $g - BR_0^{\omega} v$ is orthogonal to a certain finite set $\varphi_1, \ldots, \varphi_n$ in $C^{\infty}(G)$. Recalling that $v = CE_k u$, we thus obtain the necessary and sufficient conditions that (u,g) be in the range of (A,B):

$$u \perp N_0(A^*) \tag{8}$$

and

$$(g,\varphi_j) - (u, E_k^* C^* R_o^{\omega *} B^* \varphi_j) = 0. \tag{9}$$

From the definition (1) of E_k, it follows that

$$E_k^* v(x,t) = v(x,t) + \sum a_j \varphi(-t/j) v(x,-t/j)/j \ , \ t>0. \tag{10}$$

(We are assuming, as we may, that the measure in M^+ is the product of the measure on X and Lebesque measure on (-1,1).) By Theorem 2(i), the restri ...on to $M^{\pm} \cup X$ of $C^* R_o^{\omega *} B^* \varphi_j$ is in C^∞, hence $E_k^* C^* R_o^{\omega *} B^* \varphi_j$ is C^∞ in $M^+ \cup X$.

Notice that the inner product in (9) involves an integration over M^+. This makes sense under the hypothesis $s \geq \omega$, for then u is in $H^{s-\omega} \subset H^0$. (Actually, the whole discussion carries over directly for $s > \omega - 1/2$, since multiplication by the characteristic function of M^+ is bounded on H^t for $|t| < 1/2$.

Now consider the case where B involves only the Cauchy data of order $< m$. Then we can rewrite BR_o^{ω} as $B'R_o^{m}$, and the function appearing in (9) becomes $E_k^* C^* R_o^{m*} B'^* \varphi_j$. Since C^* has order $-\omega$ and R_o^{m*} is bounded from $\sum_0^{m-1} H^{j+1/2-t}$ to H^{-t} for $t > m - 1/2$, and the φ_j are in C^∞, we find that $\psi_j = C^* R_o^{m*} B'^* \varphi_j$ is in $H^{\omega-t}(M)$ for all $t > m - 1/2$. Since the map $\varepsilon \longrightarrow R_\varepsilon^{\omega-m} \psi_j$ is continuous into $\sum_0^{\omega-m-1/2} H^{\omega-t-j-1/2}$ for $m+1/2 - t > 0$, we find on taking $m - 1/2 < t < m+1/2$ that the normal derivatives of ψ_j of order $< \omega-m$ taken from $t > 0$ agree with those taken from $t < 0$. (Recall that $C^* R_o^{\omega*} B^* \varphi_j$ is C^∞ in $M^{\pm} \cup X$, but there is in general no agreement between the boundary values taken from the two sides of X; this is what is special in the present case $m < \omega$.) From (10) and the conditions on the coefficients a_j in formula (1) defining E_k, it follows that $E_k^* \psi_j = 0$ in M^-, the resulting function will be in $H^{\omega-m}$ (actually, in $H^{\omega-m+\varepsilon}$ for all $\varepsilon < 1/2$), and the integral $\int_{M^+} u\, E_k^* \psi_j$ makes sense for u in $H^{m-\omega}$. This is the interpretation of the orthogonality relation (9) in case $m < \omega$.

Similar arguments establish the following standard regularity result:

<u>Theorem 4.</u> If A is elliptic of degree ω, u is in $H^t(E_{M^+})$, and Au is in $H^{s-\omega}(F_{M^+})$ $(s > m-1/2)$ then $R_0^m u = \lim_{\varepsilon \to 0+} R_\varepsilon^m u$ exists in $\sum_0^{m-1} H^{t-j-1/2}$

If (A,B) is an elliptic boundary system, with B involving only Cauchy data of order $< m$, and $BR_0^\omega u = B'R_0^m u$ is in $H^{s-\omega+1/2}(G)$, $s > m-1/2$, then u is in $H^s(E_{M^+})$, and there is a constant C_{st} such that

$$||u||_s \leq C_{st}(||Au||_{s-\omega} + ||BR_0^\omega u||_{H^{s-\omega+1/2}(G)} + ||u||_t) \qquad (11)$$

<u>Proof.</u> We may assume $t<s$. Au is automatically orthogonal to $N_0(A^*)$, so by Theorem 2 there is a v in H^s with Au = Av, and $||v||_s \leq C||Au||_{s-\omega}$. When $s > m-1/2$, then $\lim_{\varepsilon \to 0+} R_\varepsilon^m v$ exists in $\sum_0^{m-1} H^{s-j+1/2}$, and by Theorem 1

$\lim_{\varepsilon \to 0+} R_\varepsilon^m (u-v)$ exists in $\Sigma H^{t-j+1/2}$, which proves the first part of Theorem 4. (Theorem 1 gives $\lim R_\varepsilon^m(u-v)$ for $m \leq \omega$, and this in turn implies the same result for $m > \omega$; the equations $A(u-v) = 0$ allows us to solve successively for the normal derivatives of u-v of order $\omega, \omega+1, \ldots$)

For the a priori inequality (11), setting $w=u-v$ as above reduces us to the case where w is in H^t, $Aw=0$, $R_0^\omega w$ is in $\Sigma H^{t-j-1/2}$, $BR_0^\omega w = B'R_0^m u - B'R_0^m v$ is in $H^{s-\omega+1/2}(G)$. Subtracting from w its projection on $N_0(A^*)$ does not change Aw or $BR_0^\omega w$, and we are reduced (by Theorem 1) to the case where $w=PR_0^\omega w$, $R_0^\omega = R^+R_0^\omega w$. From Corollary 1 above, $||P^+R_0^\omega w||_{s-\frac{1}{2}} \leq C_{st}(||BP^+R_0^\omega w||_{s-\omega+1/2} + ||R_0^\omega w||_{t-\frac{1}{2}})$, and Theorem 4 follows from the continuity of the surface potential P.

So far we have considered the pair (A,B) as a bounded operator from $H^s(E_{M^+})$ to $H^{s-\omega}(E_{M^+}) \oplus H^{s-\omega-1/2}(G)$. Another common way to treat boundary problems uses the boundary operator B to define the domain of an unbounded operator A_B on $H^0(E_{M^+})$.

<u>Definition 2.</u> Let A be a differential operator of order ω on E_{M^+}, and B a ψdo: $\Sigma H^{s-j-1/2}(E_X) \longrightarrow H^{s-\omega+1/2}(G)$. Then A_B is the unbounded operator A acting on the domain $\{u \text{ in } H^\omega(E_{M^+})\text{: } BR_0^\omega u = 0\}$. From Theorems 3 and 4 we obtain immediately:

Theorem 5. If (A,B) is an elliptic system, then A_B is closed, it has a C^∞ finite-dimensional null space, and the range is the orthogonal complement of finitely many C^∞ functions.

Recall that Definition 1 requires the top-order symbol b_o of B to be surjective; this is natural in studying the range of the map $u \longrightarrow (Au, BR_o^\omega u)$, but not so natural for the operator A_B, where it is only the null space of B that matters. For this latter form of boundary problems we can widen the theory by replacing B with a map acting in the space of Cauchy data. To set things up conveniently in this context, we modify the Cauchy data so that all the entries in the boundary operator can have degree zero. Set $R'u = (\Lambda^{\omega-1}u(0), \Lambda^{\omega-2}D_t u(0),\ldots,D_t^{\omega-1}u(0))$, where Λ is an invertible elliptic ψdo_1 on $C^\infty(E_X)$. Similarly, replace $P^+ = (P_{jk}^+)$ by $P^{+'} = (\Lambda^{\omega-j}P_{jk}^+ \Lambda^{k-\omega})$ $(1 \leq j,\ k \leq \omega)$ and $P = (P_1,\ldots,P_\omega)$ by $P' = (P_1\Lambda^{1-\omega},\ldots,P_\omega)$. Then the "adjusted Cauchy data" $R'u$ map H^s into $\sum_0^{\omega-1} H^{s-\omega-1/2}$, and $P'R'u = PRu$, and $P^{+'}$ acting in the adjusted Cauchy data is equivalent to P^+ acting in the standard Cauchy data. Moreover, given a boundary operator $B = (B_0,\ldots,B_{\omega-1})$ acting on the standard Cauchy data, the operator $B' = (B_0\Lambda^{1-\omega},\ldots,B_{\omega-1})$ satisfies $B'R'u = BR_0^{\omega}u$; and (A,B) is elliptic in the sense of Definition 1 if and only if $\sigma_0(B)(\xi_x)$ maps the range of $\sigma_0(P^{+'})(\xi_x)$ isomorphically onto the fibre G_x. To extend the definition of ellipticity, this surjectivity condition is weakened.

Definition 3. Let A be an elliptic system of order ω, and let B be a ψdo_0 mapping $\sum_0^{\omega-1} H^s(E_X)$ into itself for each s. B is well-posed for A if (i) the range of B is closed for each s, and (ii) $\sigma_0(B)(\xi_x)$ maps the range of $\sigma_0(P^{+1})(\xi_x)$ isomorphically onto the range.

Now we define A_B just as before. It is easy to prove:

Lemma 2. Let (A,B) be an elliptic system in the sense of Definition 1. Then there is a ψdo_0, $\tilde{B}$, such that $\tilde{B}$ is well-posed for A, and

$\tilde{B}(\Lambda^{\omega-1}g_0,\ldots,\Lambda g_{\omega-2},g_{\omega-1}) = 0$ if and only if $B(g_0,\ldots,g_{\omega-2}, g_{\omega-1}) = 0$.

Lemma 3. Let B be well-posed for A, as in Definition 3. Then there is a projection B', a ψdo_0, such that B and B' have the same null space, and B' is well-posed for A.

Lemma 4. Let B be well-posed for A, and assume B is a projection. Then there is a ψdo_0, call it D, such that $DBP^+ - P^+$ and $BP^+D - B$ have degree $-\infty$.

Theorem 6. If B is well-posed for A, then the conclusions of Theorems 4 and 5 hold.

Lemma 2 shows that the new class of problem includes the old, and Theorem 6 shows that the new class enjoys the same properties. The simplest example showing that the new class actually extends the old is found by taking M^+ to be the unit disk in R^2, $A = \partial/\partial x + i\partial/\partial y$. Then X is the unit circle, $T'(X)$ is the disjoint union of two half-infinite circular cylinders; the "adjusted" Cauchy data and the usual Cauchy data coincide (since $\omega=1$), and $\sigma_0(p^+)(\xi_x) = 1$ on the "upper" half-infinite cylinder, and $=0$ on the "lower" half-infinite cylinder. Hence there is no elliptic system (A,B) in the sense of Definition 1 (since $\dim G_x = 1$ and $\dim G_x = 0$ are incompatible). On the other hand, there is a well-posed operator B in the sense of Definition 3; in fact, for any elliptic A, $B = P^{+\prime}$ is well-posed for A. In the present example, $P^+ = P^{+\prime}$ is essentially the Hilbert transform.

To prove Lemma 2, we pass from B to $B' = (B_0\Lambda^{1-\omega},\dots,B_{\omega-1})$ as above. Then the null-space of B' on H^s is the orthogonal complement of the range of B'^* on H^{-s}; by Theorem IV. 7 , we have an orthonormal projection on this range, and the complementary projection gives the desired $\widetilde{B}$.

The proof of Lemma 3 imitates part of the proof of Theorem IV. 7 , Lemma 4 is proved like Lemma 1, and then Theorem 6 follows the lines of Theorems 4 and 5.

We have pointed out one advantage of Definition 2 over Definition 1; for any elliptic A, there is a well-posed boundary operator B. Another advantage of this wider class is that it is closed under the taking of adjoints. The proof of this depends on the "Green's formula" for A,

$$(A^{*}u,v) - (u,Av) = (R'u, \cdot' R'v) \qquad (12)$$

where $\mathfrak{A}'$ is a triangular invertible matrix of ψdo's of degree $1-\omega$, and R' is the adjusted Cauchy data. Formula (12) is obtained easily by writing $A = \Sigma A_j D_t^{\omega-j}$ and integrating by parts with respect to t. ($\mathfrak{A}'$ is related to the operator $\mathfrak{A}$ in Seeley [5] by $\mathfrak{A}' = (\mathfrak{A}'_{jk}) = (\Lambda^{j-\omega}\mathfrak{A}_{jk}\Lambda^{k-\omega})$, $1 \leq j,k \leq \omega$.)

<u>Theorem 7.</u> If B is well-posed for A, and B is a projection, then $(A_B)^{*} = (A^{*})_C$, where A^{*} is the formal adjoint of A, and

$$C = \Lambda^{\omega-1}(I - B^{*})\mathfrak{A}'^{*} \qquad (13)$$

is well-posed for A^{*}.

(The adjoint B^* is taken with the inner product of $\sum_0^{\omega-1} H^0(E_X)$. By Lemma 3, the hypothesis that B is a projection does not restrict the generality.)

<u>Proof.</u> For v in the domain of A_B, $R'v = (I-B)R'v$, so by (12)

$$(u,Av) = (A^*u,v) - ((I-B^*)\mathfrak{A}'^*R'v), \ v \text{ in domain } (A_B). \qquad (14)$$

Hence for u in the domain of A_C^*, $|(u,Av)| = |(A^*u,v)| \leq c_u||v||_o$, so u is in the domain of $(A_B)^*$, by definition.

Suppose, conversely, that u is in the domain of $(A_B)^*$. Then u and Au are in H^0, so by Theorem 4, $R'u$ exists in $\sum_0^{\omega-1} H^{1/2-\omega}(E_X)$. We will show that $(I-B^*)\mathfrak{A}'^*R'u = 0$, as follows. Given f in $\sum_0^{\omega-1} C^\infty(E_X)$, take v in $C^\infty(E)$ such that $R'v = (I-B)f$. Since B is a projection, v is in the domain of A_B, and so is φv if $\varphi \equiv 1$ near X. Choose a sequence φ_n of such functions, φ_n tending boundedly to zero in M^+. Then

$$(u,A\varphi_n v) = (A^*u,\varphi_n v) - ((I-B^*)\mathfrak{A}'^*R'u,\ (I-B)f)$$

$$= (A^*u,\varphi_n v) - ((I-B^*)\mathfrak{A}'^*R'u,f) \ ,$$

since B is a projection. Since u is in the domain of $(A_B)^*$, we get $|(u,A\varphi_n v)| \leq c||\varphi_n v||_0 \longrightarrow 0$, hence $((I-B^*)\mathfrak{a}'^* R'u,f) = 0$. Since this holds for all f in C^∞, we find $CR'u = 0$, as desired.

Finally, we note that $\mathfrak{a}'$ is invertible from $\Sigma\, H^s(E_X)$ to $\Sigma\, H^{s-\omega+1}(E_X)$ and B is a projection, so $C = \Lambda^{\omega-1}(I-B^*)\,\mathfrak{a}'^*$ has closed range as an operator in $\Sigma\, H^s(E_X)$. Further, recalling the characterization of the top symbol $\sigma_0(P^{+'})$ in terms of ordinary differential equations, and repeating the above arguments when A is a system of constant coefficient ordinary differential equations on the half line, one can deduce that $\Lambda^{\omega-1}(I-B^*)\,\mathfrak{a}'^*$ is well-posed for A^*. Hence, by what we have shown already, the regularity part of Theorem 6 guarantees that the domain of $(A_B)^*$ lies in H^ω, and the proof is complete.

As a corollary of the proof, we obtain

<u>Theorem σ.</u> If B is well-posed for A, and if A_B^0 denotes the operator A acting on the domain

$$\{u \text{ in } C^\infty(E_{M^+ \cup X}) : BR'u = 0\},$$

then A_B is the closure of A_B^0.

Proof. We may take B to be a projection. Then the proof of Theorem 7 shows that $(A_B^0)^* = A_C^*$, and a similar argument shows that $(A_C^*)^* = A_B$. Since in general T^{**} is the closure of T, the proof is complete.

The Dirichlet problem is an important special case of Theorem 7. In that case ω is even, and B is the projection on the first $\omega/2$ entries in the adjusted Cauchy date. Hence $B^* = B$, and $I-B^*$ is the projection on the last $\omega/2$ entries. From the triangular nature of $\mathcal{Q}'$ (notice that the "hypotenuse" of the triangle is the secondary diagonal, not the main diagonal, so $\mathcal{Q}'$ tends to reverse the order of the Cauchy data), it follows that $(I-B^*)\mathcal{Q}'^*f = 0$ if and only if $Bf = 0$. Hence: if the Dirichlet data B are well-posed for A, then $(A_B)^* = A^*_B$. Similar remarks apply to the Neumann problem, where B projects on the last $\omega/2$ entries in the Cauchy data.

Another way to work with the Dirichlet problem is to consider A as an operator from $H_0^{\omega/2}(E_{M^+})$ to $H^{-\omega/2}(E_{M^+})$; the point of this is that each of these spaces is the dual of the other with respect to the pairing $\varphi, \psi \longrightarrow \int_{M^+} \varphi\psi$.

Theorem 9. If the Dirichlet boundary conditions are well-posed for A, then

$$A: H_0^{\omega/2}(E_{M^+}) \longrightarrow H^{-\omega/2}(F_{M^+}) \tag{15}$$

has C^∞ finite-dimensional null space, and the range is the orthogonal complement of finitely many C^∞ functions with vanishing Cauchy data of order $< \omega/2$. The adjoint of the map in (15) with respect to the pairing of $H_0^{\omega/2}$ and $H^{-\omega/2}$ is given by the formal adjoint

$$A^*: H_0^{\omega/2}(F_{M^+}) \longrightarrow H^{-\omega/2}(E_{M^+}).$$

<u>Proof.</u> Here C^∞ means C^∞ on $M^+ \cup X$. Everything but the statement about A^* follows from Theorem 3; for the vanishing Cauchy data, see the remarks at the end of that proof. The statement about A^* follows from the way $H_0^{\omega/2}(M_+)$ and $H^{-\omega/2}(M_+)$ are identified as duals (or antiduals), and from the fact that functions vanishing near X are dense in $H_0^{\omega-2}$.

<u>Remark.</u> The way we viewed Dirichlet boundary conditions as a projection is closely related to the idea of boundary operators of "normal" type: $B = (B_1, \ldots, B_{\omega/2})$ is of normal type if there is a $C = (C_1, \ldots, C_{\omega/2})$ such that the Cauchy data of order $< \omega$ are determined uniquely and continuously by BR_0^ω and CR_0^ω together. In this case, we can replace the Cauchy data $R_0^\omega u$ in all

our arguments by $(BR_0^{(\omega)}u, CR_0^{(\omega)}u)$; only the details of the boundary form ' in Green's formula (12) are changed. The fact that B is a projection (on the first half of $(BR_0^{(\omega)}, CR_0^{(\omega)})$) is exploited more or less as in the proof of Theorem 7.

Many variations of boundary problems have been considered. For instance, Lions and Magenes [1] have studied (A,B) and A_B on spaces such as

$$H_A^{t,s} = \{u \text{ in } H^t;\ Au \in H^{s-\omega}\}$$

with norm

$$|||u|||_{t,s} = (||u||_t^2 + ||Au||_s^2)^{1/2} .$$

<u>Theorem 10.</u> If B involves only Cauchy data of order < k, then (A,B) is continuous from $H_A^{t,s}$ to $H^s(E_{M^+}) \oplus \Sigma H^{t-\omega_j-1/2}(E_X)$ for $s \geq k$, and the other conclusions of Theorem 3 hold.

<u>Proof.</u> As in the proof of Theorem 4, we find that $\lim_{\varepsilon \to 0+} BR_\varepsilon^{(\omega)}u$ exists in $\Sigma H^{t-\omega_j-1/2}$ and is dominated by $|||u|||_{t,s}$, so (A,B) is bounded. The null space and range are analyzed precisely as in the proof of Theorem 3.

A slightly different approach is taken by Beals [1], who considers the domain

$$\ldots = \{u \text{ in } H^0(M_+): Au \,\varepsilon\, H^0(M_+),\ R_0^{\omega} u \,\varepsilon \sum_0^{\omega-1} H^{\omega-1-j}(X)\}.$$

Appealing to Theorems 1 and 2, we write

$$u = CE_k Au + PR_0^{\omega}(u - CE_k Au) + u_0 ,$$

where u_0 is the orthogonal projection of u on $N_0(A)$. Since $CE_k Au$ is in H^{ω} and $R_0^{\omega} u$ in $\Sigma H^{\omega-1-j}$, it follows that u is in $H^{\omega-1/2}$, and in fact ... coincides with the $H_A^{\omega-1/2,\omega}$ considered above. The same argument shows in general, for $t \leq s$ and $s > \omega - 1/2$, that elements u in $H_A^{t,s}$ are characterized by Au and $R_0^{\omega} u$:

$$H_A^{t,s} = \{u \text{ in } \cup H^k: Au \,\varepsilon\, H^{s-\omega},\ \lim_{\varepsilon \to 0+} R_\varepsilon^{\omega} u \text{ exists in } \Sigma H^{t-j-1/2}\}.$$

Appendix

The proofs of Theorem 1 and 2 are easiest in case the given operator A is elliptic and invertible on the whole of the compact unbounded manifold M. In the appendix we show how to pass from that case to the general case, correcting errors in our paper [5].

Given $A: C^\infty(E_{M^+}) \longrightarrow C^\infty(F_{M^+})$, form the system

$$D = \begin{pmatrix} 0 & A^* \\ A & 0 \end{pmatrix}, \qquad (16)$$

which is formally self-adjoint and elliptic on $C^\infty(E_{M^+} \oplus F_{M^+})$. Let $N_0(D) = \{u \text{ in } C^\infty: u = 0 \text{ in } M^-,\ Du = 0 \text{ in } M^+\}$, and let P_0 be the orthogonal projection on $N_0(D)$.

Lemma. There is an invertible elliptic ψdo_ω, $\tilde{A}$, on $C^\infty(E \oplus F)$ such that in a neighborhood of $M^+ \cup X$, $\tilde{A}u = Du + P_0 u$.

Proof. Let $d = \sigma_\omega(D)$ be the top symbol of D. Since D is self-adjoint, d has real eigenvalues, and there is a C^∞ homotopy of d to the identity, say

$$d(\theta) = \frac{i}{2\pi} \int_\Gamma \lambda^\theta (d-\lambda)^{-1} d\lambda, \quad 0 \leq \theta \leq 1 ,$$

where Γ is a path surrounding the eigenvalues of d but not crossing the cut $\{\lambda: \lambda \geq 0\}$. With such a homotopy we can extend d from $M^+ \cup X$ to all of M in such a way that d is always non-singular and homogeneous of degree ω, is an elliptic polynomial in a neighborhood of $M^+ \cup X$, and $d(\xi_x)$ is a scalar multiple of the identity for all x outside some neighborhood of $M^+ \cup X$. (The sole purpose of this last condition is to guarantee that the extension can be made to all of M). Call this extended symbol d', and let D' be an operator with $\sigma_\omega(D') = d'$. Let $\varphi^2 + \psi^2 = 1$, where $\psi = 0$ in a neighborhood of $M^+ \cup X$, and $\psi = 1$ outside another slightly larger neighborhood of $M^+ \cup X$; then set

$$D_1 = \varphi D \varphi + \psi D' \psi .$$

(We assume, as we may, that A, and hence D, is defined in a neighborhood of X). The operator D_1 is elliptic, so its null space is C^∞ and finite-dimensional. The subspace $N_o(D_1)$ of the null space having support in $M^+ \cup X$ is thus finite-dimensional, and since $N_o(D_1) = N_o(D) = N_o(A) \dotplus N_o(A^*)$, all these spaces are finite dimensional. It follows that the projection P_o on $N_o(D_1)$ is a $\psi do_{-\infty}$.

We now modify D_1 to make it invertible. Let $D_2 = D_1 + P_o$; then

$$\text{no function in the null space } N(D_2) \text{ has support in } M^+ \cup X. \tag{17}$$

Since $\sigma_\omega(D_2)$ is homotopic to a scalar multiple of the identity, the index of D_2 is zero, and the null spaces $N(D_2)$ and $N(D_2^*)$ have the same dimension, call it n. We make D_2 invertible by adding to it an appropriate map taking $N(D_2)$ onto a subspace of $C_c^\infty(E_{M^-})$.

By condition (17), there are n-dimensional subspaces S and S^* of $C_c^\infty(E_{M^-})$ such that the map $\varphi \longrightarrow (\cdot, \varphi)$ is an antiisomorphism of S onto the dual of $N(D_2)$, and of S^* onto the dual of $N(D_2^*)$. Because $N(D_2^*)$ is orthogonal to the range of D_2, S^* and the range of D_2 are linearly independent. Let $\varphi_1, \ldots, \varphi_n$ and $\varphi_1^*, \ldots, \varphi_n^*$ be orthonormal bases of S and S^*, and set

$$\tilde{A}u = D_2 u + \Sigma\,(u, \varphi_j)\varphi_j^* .$$

This A is invertible, for its null space is trivial and its index is zero; thus the lemma is proved.

Now assume Theorem 1 is proved for an invertible differential operator A. Given a general A, form $\tilde{A}$ as above. This is no longer a differential operator, but near X it differs from a differential operator only by the projection P_o on a finite-dimensional space of C^∞ functions vanishing in M^-. The construction of potentials and projections applies equally well in this situation, and hence we have for $\tilde{A}$ a potential $\tilde{P}$ and projections $\tilde{P}^\pm$. Since $\tilde{A}$ occurs naturally as a 2×2 matrix $(\tilde{A}_{ij})$, we can write $\tilde{P} = (\tilde{P}_{ij})$, $\tilde{P}^\pm = (\tilde{P}^\pm_{ij})$. Define

$$P = (\tilde{P}_{11}), \quad P^\pm = \tilde{P}^\pm_{11} .$$

Since $\tilde{P}$ maps $\Sigma H^{s-j-1/2}$ into H^s, the same is true of P. Since $\tilde{A}\tilde{P} = 0$ in M^+, we have $AP = 0$ in M^+, and the range of P is orthogonal to $N_o(A)$. Conversely, if u is in the null space $N(A,s)$ of A acting on $H^s(E_{M^+})$, and u is orthogonal to $N_o(A)$, then $(u,0)$ is in $N(\tilde{A},s)$, hence $(u,0) = \tilde{P}R_o^{(\omega)}(u,0)$, hence $u = \tilde{P}_{11}R_o^{(\omega)}u = PR_o^{(\omega)}u$. Thus parts (i) - (iii) of Theorem 1 are established, and part (iv) is immediate. For part (v), since $Pf \in N(A,s)$, it follows that $P^+f = \lim R_\varepsilon^{(\omega)}Pf$ lies in $R_o(A,s)$. Conversely, if $f = R_o^{(\omega)}v$ lies in $R_o(A,s)$, we can let u be the projection of v on the orthogonal complement of $N_o(A)$; then $R_o^{(\omega)}u = R_o^{(\omega)}v$, and as we showed above, $u = PR_o^{(\omega)}u$, so $u = Pf$ and $P^+f = R_o^{(\omega)}u = f$, i.e. f lies in the range of P^+. This completes the proof of Theorem 1 in the general case.

To prove Theorem 2, we write $\tilde{A}^{-1}$ as a matrix (C_{ij}), and take $C = C_{12}$; in other words, if $\tilde{A}(v,w) = (0,u)$, then $Cu = v$. The first part of Theorem 2, stating that $C^* R_o^{\omega*}|_{M^\pm}$ is continuous from $\Sigma H^{j-s+1/2}(F_X)$ to $H^{\omega-s}(E_{M^+})$ for each s, follows immediately from the continuity properties established in the proof of Theorem 1 in [5]. The second part of Theorem 2 states that for u in $H^s(F_{M^+})$ and orthogonal to $N_o(A^*)$, we have $ACE_k u = u$ in M^+, where E_k is an extension map: $H^s(E_{M^+}) \longrightarrow H^s(E)$, and $k \geq |s|$. To prove this recall that in M^+ we have

$$\tilde{A} = \begin{pmatrix} 0 & A^* \\ A & 0 \end{pmatrix} + P_o ,$$

where P_o is orthogonal projection on

$$N_o\left(\begin{pmatrix} 0 & A^* \\ A & 0 \end{pmatrix}\right) = N_o(A) \oplus N_o(A^*).$$

If u is orthogonal to $N_o(A^*)$, so is $E_k u$, and the second component of $P_o(v,w)$ vanishes when $\tilde{A}(v,w) = (0,u)$, hence $ACE_k u = E_k u = u$ in M^+, and Theorem 2 is proved.

R. Seeley

References

S. Agmon

[1] "On the eigenfunctions ..." Comm. Pure Appl. Math 15 (1962) 119-147.

M. F. Atiyah and I. M. Singer

[1] "The Index of elliptic operators on compact manifolds". Bull. Amer. Soc. 69 (1963) 422-433.

R. Beals

[1] "Non-local boundary value problems for elliptic operators". Amer. J. I 87 (1965) 315-362.

M. Berger

[1] Le spectre des variétés riemanniennes"(to appear in Romanian journa

L. Boutet de Monvel ....

T. Burak

[1] "Fractional powers of elliptic differential operators". Annali della Scuola Normale Sup. di Pisa 22 (1968) 113-132.

A. P. Calderon and A. Zygmund

[1] "On the theorem of Hausdorff-Young and its extensions", Annals of Math. Studies No. 25, pp 166-188, Princeton University Press, 1950.

T. Carleman

[1] "Properties asymptotiques ..." C. R. 8eme Congr. des Math. Scand. Stockholm 1934 (Lund 1935) 34-44.

D. Fujiwara

[1] "On a special class of pseudo-differential operators", Journ. Fac. Sci. Univ. of Tokyo 14 Part 2 (1967) pp 221-249.

[2] "On the asymptotic formula for the Green operators of elliptic operators on compact manifolds", same journal, pp 251-283.

I. C. Gohberg

[1] "On the theory of multidimensional singular integral equations," Soviet Math Dokl. 2 (1961) 960-963.

P. Greiner

[1] "On zeta functions connected with elliptic differential operators" University of Toronto preprint.

[2] "Asymptotic expansions for the heat equation", lecture S5A, Conference on Global Analysis, Berkeley, July 1968.

L. Hörmander

[1] Linear Partial Differential Operators, Academic Press, New York 1963.

[2] "Pseudo-differential operators", Comm. Pure Appl. Math. 18 (1965) 269-305.

[3] "Paeudo-differential operators and hypo-elliptic equations," AMS Symposia in Pure Math. vol. 10.

[4] "Pseudo-differential operators and non-elliptic boundary problems". Annals of Math 83 (1966) 129-209.

J. J. Kohn and L. Nirenberg

[1] "An algebra of pseudo-differential operators", Comm. Pure Appl. Math. 18 (1965) 501-517.

[2] "Non-coercive boundary problems," same journal, 443-492.

'. Kree

[1] "Les noyaux des operateurs pseudo-differentials," Publications du séminaire de mathematiques de l'Université de Bari.

L. L. Lions and E. Magenes,

[1] Problèmes aux limites non homogènes et applications, vol. I Ed. Dunod. Paris 1968.

H. P. McKean and I. M. Singer

[1] "Curvature and the eigenvalues of the Laplacian," Journ. Diff. Geom. 1 (1967).

L. Nirenberg [1]

M. Riesz

[1] "L'Integrale de Riemann-Liouville ..." Acta Math. 81 (1949) 1-222.

F. Riesz and B. Sz. Nagy

[1] Lecons d'Analyse Fonctionnelle.

R. Seeley

[1] "Singular integrals on compact manifolds", Amer. Journ. Math. 81 (1959) 658-690.

[2] "Complex Powers of an elliptic operator", Amer. Math. Soc. Proc. Symp. Pure Math. 10, 288-307.

[3] "The resolvent of an elliptic boundary problem", to appear in Amer. J. Math.

[4] "Analytic extension of the trace associated with elliptic boundary problems", to appear in Amer. J. Math.

[5] "Singular integrals and boundary value problems", Amer Journ. Math. 80 (1966) 781-809.

H. Weyl

[1] "Das asymptotische ..." Rend. Circ. Mat. Palermo 39 (1915), 1-50.

CENTRO INTERNAZIONALE MATEMATICO ESTIVO

(C.I.M.E.)

E. SHAMIR

"BOUNDARY VALUE PROBLEMS FOR ELLIPTIC CONVOLUTIONS SYSTEMS"

Corso tenuto a Stresa dal 26 Agosto al 3 Settembre 1968

BOUNDARY VALUE PROBLEMS FOR ELLIPTIC CONVOLUTION SYSTEMS

by

Eliahu Shamir

(University of Jerusalem)

0. Introduction and notations.

Whe shall study here generalized boundary-value problems (including potentials) for homogeneous elliptic systems of convolution equations in a half-space. The main result is Theorem 3, giving the necessary and sufficient conditions for a problem to be well-posed. Actually we establish an isomorphism between the space of unknowns and the data space. We use a pure L^2-theory for systems of order 0. But we outline the way to obtain L^p - theory and also the way to treat systems with various degrees of homogeneity - by introducing suitable H^s spaces for the various components. At the end we add several other remarks. More details will appear in [8], which is a natural continuation of [7].

As in the special case of elliptic differential systems (cf. [1], [5]), a general pseudo-differential problem in a bounded domain has at each boundary point of the domain a "tangential" or "first approximation" problem. This is a half-space homogeneous convolution problem involving only the principal parts of the original operators. The general problem is well-posed if and only if all these tangential problems are well posed. This yields the general covering condition. Moreover the estimates for the general case are obtained from those of the tangential problems by a well known technique - using Korn's principle. Thus most of the novel ideas and difficulties in treating boundary problems for pseudo-differential systems are encountered in the half-space homogeneous case, which we treat here.

We mention though one additional difficulty which is not encountered in the half-space case, or in the general differential case. The relevant order(s) or "indices" may vary, for pseudo-differential problems along the boundary. This may necessitate the introduction of function spaces with varying degree of differentiability. We refer the reader to Visik-Eskin papers [9, 10] which treat general boundary value problems for pseudo-differential systems.

NOTATIONS: We define the Fourier transform by

$$u^{\wedge}(\eta) = (Fu)(\eta) = \int_{-\infty}^{\infty} u(y) e^{+iy.\eta} dy .$$

Similarly in n-dimensional space R^n. $R_+ = \{y, y \geq 0\}$ C = the complex field. $C_+ = \{\text{complex } \eta, \ Im > 0\}$. Similarly for R_-, C_-. Note that:

u(y) is supported in $R_{\pm} \Longleftrightarrow u^{\wedge}(\eta)$ extends holomorphically to $C_{\pm}$. $Y_{\pm}$ denotes the characteristic function of $R_{\pm}$ and also, in $L^2(R)$, the operator of mutiplication by that function. Since Y_+ is used more often that Y_- we usually use Y instead of Y_+.

$Y^F_{\pm} = FY_{\pm}F^{-1}$, i.e. the operator $Y_{\pm}$ induces in FL^2. It is the projection on the subspace of functions having holomorphic extension to $C_{\pm}$.

In Section 3 we extend the notation in an obvious fashion to $R^n = (x_1, \ldots, x_{n-1}, y) = (x, y)$ with dual variables (ξ, η).

In the remarks of the last section and at few other places

we mention H^S spaces, but their use in this paper is not essential.

1. Scalar one-dimensional problems.

We start by studying the one-dimensional operator

(1.1) $$A^k = YF^{-1}(\eta+i)^k(\eta-i)^{-k}FY$$

acting in YL^2. Here k is an integer.

LEMMA 1.1. If $k \geq 0$ then A^k is onto and dimker $A^k = k$. The space Ker A^k is spanned by

(1.2) $$F^{-1}(b_r(\eta)) = F^{-1}(\eta^r(\eta+i)^{-k}), \quad 0 \leq r < k.$$

For arbitrary $Yf \in YL^2$ and complex numbers $\Phi_r, 0 \leq r < k$, there is a unique solution $Yu \in YL^2$ satisfying

(1.3) $$A^k u = Yf$$

(1.4) $$F^{-1}[\eta^r(\eta-i)^{-k} \cdot (Yu)^\wedge(\eta)](0) = \Phi_r, \quad 0 \leq r < k \ ;$$

and there is a norm equivalence

$$||Yu|| \sim ||Yf|| + \Sigma|\Phi_r|.$$

For $\psi_r = 0$, the unique solution is given by $Yu = A^{-k}f$.

LEMMA 1.2. If $k \leq 0$, A^k is 1-1 (surjective) with closed range, and dimCoker $A^k = |k|$. A complementary subspace to Range A^k is spanned by

(1.5) $\quad F^{-1}(p_r(\eta)) = F^{-1}(\eta^r(\eta+i)^k), \quad 0 \le r < |k|.$

There is a unique function $Yu \in YL^2$ and a unique $|k|$-tuple of numbers $(\psi_0, \dots, \psi_{|k|-1})$ satisfying

(1.6) $\quad A^k u = Yf - Y \sum_{r=0}^{k-1} \psi_r F^{-1}(\eta^r(\eta+i)^k), \quad Yf$ given in YL^2,

and there is a norm-equivalence

$$\|Yu\| + \sum |\psi_r| \sim \|Yf\|.$$

PROOFS. Verification of the first two statements in each lemma is direct It is useful though to observe the effect of changing the (complex) variable η to z by the standard conformal map taking C_+ to the unit disc $|z|<1$: $z=(\eta-i)(\eta+i)^{-1}$. Obviously A^k goes over to the operator $Pz^{-k}P$ where P is the natural projection in the space of L^2 - Fourier series

$$P \sum_{n=-\infty}^{\infty} a_n z^n = \sum_{n=0}^{\infty} a_n z^n, \quad |z|=1.$$

Now the functions $1, z, \dots, z^{|k|-1}$ span the kernel if $k>0$, the cokernel if $k<0$.

Next we notice that the boundary condition (1.4) can be written as a scalar product in FYL^2

(1.7) $\quad \frac{\Phi_r}{2\pi} = \int_{-\infty}^{\infty} \eta^r(\eta-i)^{-k}(Yu)^\wedge(\eta)d\eta = \langle \eta^r(\eta+i)^{-k}, (Yu)^\wedge(\eta) \rangle.$

The left factor in the scalar product $\langle , \rangle$ is analytic in C_+, hence belongs to FYL^2. In the general, if the symbol of a boundary condition is $B_r(\eta)$ but $\overline{B}_r(\eta)$ (complex conjugate) is not in FYL^2, we have to take its projection on FYL^2, i.e. the boundary condition has the form

$$(1.8) \qquad \langle Y^F \dot{\overline{B}}_r(\eta) \; , \; (Yu)\hat{\ }(\eta) \rangle = \frac{\Phi_r}{2\pi} \; .$$

Now it is well known that there is a unique solution of an operator equation ($A^k u = YF$ in our case) which is orthogonal to the kernel of that operator (this is problem (1.3) - (1.4) with $\Phi_r = 0$), or more generally which has a given orthogonal projection on the kernel (this is the general case).

One can verify directly the last statement of Lemma 1.1, but it is illuminating to use the following argument. Change the unknown u to v by the relation $v\hat{\ }(\eta) = (\eta - i)^{-k} u\hat{\ }(\eta)$. This operator maps L^2 onto H^k, but preserves supports in R_-, hence also maps YL^2 onto

$$(1.9) \qquad H^k(R_+) = H^k/H^k_- = YH^k \, , \quad (u \in H^k_{\pm} \Longleftrightarrow \text{Support } (u) \subset R_{\pm}) .$$

The equations (1.3)-(1.4) for v are

$$(1.10) \qquad YF^{-1}\left[(\eta+i)^k (Yv)\hat{\ }(\eta)\right] = Yf, \quad F^{-1}\left[\eta^r (Yv)\hat{\ }(\eta)\right](0) = 0, \quad 0 \leq r < k.$$

This is a Dirichlet problem. Since $F^{-1}(\eta+i)^{\pm k} F$ preserves the support in R_+, it is immediately seen that $Yv = YF^{-1}\left[(\eta+i)^{-k} (Yf)\hat{\ }(\eta)\right]$

is the (unique) solution. It belongs to H_+^k and so has vanishing Dirichlet data. Returning to the original u, we get $Yu = A^{-k}f$.

Finally, since we proved that the problem (1. 3)-(1. 4) gives an isomorphism $YL^2 \to YL^2 \times C^k$, we have the norm-equivalence claimed in Lemma 1. 1.

To solve equation (1. 6) in Lemma 1. 2, we subtract from Yf its projection on the complementary subspace spanned by (1. 5). (1. 5). This determines the ψ_r uniquely, and the right-hand side of (1. 6) now belongs to range A^k. Then Yu is uniquely determined. Actually, when Yg is in Range A^k ita (unique) inverse image is given by $Yu = A^{-k}g$. Thus we have established an isomorphism

(1. 11) $$YL^2 \times C^{|k|} \to YL^2 ;$$

(1. 12) $$(Yu, \psi_0, \dots, \psi_{|k|-1}) \to A^k u + Y\sum_r \psi_r \, F^{-1}[\eta^r(\eta+i)^k]. \quad (k<0).$$

Y on the right is superfluous here since the "symbols" $\eta^r(\eta+i)^k \in FYL^2$, but will be needed for other symbols.

The generalized (or "dual") problem (1. 6) is referred to as a "problem with potentials" since the solution contains potential terms wich take the boundary values ψ_r (potential densities) into functions in the domain R_+ by means of "potential kernels" with symbols $\eta^r(\eta+i)^k$. This kind of problem is natural when the domain operator (A^k here) by itself is overdetermined (has a non-trivial cokernel), whereas boundary conditions are needed (as in Lemma 1. 1) when the operator is underdetermined (has a non-trivial kernel).

From the proof of Lemma 1.1, in particular from (1.8), it follows that a set B of k L^2-symbols for boundary conditions $B_r(\eta)$ can replace those of the Lemma if and only if

(1.13) $\quad Y^F \bar{B}_r(\eta)$ <u>span</u> $(\mathrm{Ker}\ A^k)\hat{} = \mathrm{Ker}\ (A^k)\hat{}$.

Dually, $|k|$ potential-symbols $P_r(\eta)$ will serve for Lemma 1.2 if and only if

(1.4) $\quad \{Y^F P_r(\eta)\}$ <u>span a complementary space to</u> $(\mathrm{Range}\ A^k)\hat{} = \mathrm{Range}\ (A^k)\hat{}$

(Succinctly we say they span $\mathrm{Coker}(A^k)\hat{}$, meaning that the images under the natural map $v \longrightarrow v + \mathrm{Range}(A^k)\hat{}$ span the cokernel).

A set B of P of $|k|$ symbols satisfying (1.13) if $k>0$ or (1.14) if $k<0$ is said to <u>cover</u> the operator A^k. The reason for this name is explained in Section 4.

LEMMA 1.3. For $-1/2 < \mathrm{Re}\,\zeta < 1/2$ the operator A^ζ is an isomorphism of YL^2, its inverse being $A^{-\zeta}$.

This statement is equivalent to the fact that the map $u \rightarrow (Y_- u, Y_+ u)$ of $H^s \rightarrow Y_- H^s \times Y_+ H^s$ is an isomorphism for $-1/2$ s $1/2$ we refer to [6, 7]. Also the spaces H^ζ, $Y_{\pm} H^\zeta$ depend only on $\mathrm{Re}\,\zeta$. (For $\zeta = 1/2$ the range of A^ζ is not closed.)

In view of Lemma 1.3, $\mathrm{Ker}\ A^{k+\zeta} = A^{-\zeta}\mathrm{Ker}\ A^k$ and $\mathrm{Range}\ A^{k+\zeta} = A^\zeta\ \mathrm{Range}\ A^k$. Also $A^{-\zeta} = (A^\zeta)^{-1}$ means that

A^{ζ}, $A^{-\zeta}$ commute with Y (and their symbols commute with Y^F). Thus we obtain for $A^{k+\zeta}$ the same results as for A^k. Sumarizing we have

THEOREM I. Let $-1/2 < \mathrm{Re}\,\zeta < 1/2$. For $k \geq 0$ the boundary problem

$$A^{k+\zeta} u = Yf \,, \langle Y^F \overline{B}_r(\eta), \ (Yu)\hat{}(\eta) \rangle = \frac{\Phi_r}{2\pi} \,, \quad 1 \leq r \leq k,$$

gives an isomorphism $YL^2 \to YL^2 \times C^k$ if and only if $\{Y^F \overline{B}_r\}$ span $\mathrm{Ker}(A^{k+\zeta})\hat{}$ whose dimension is k. For $k \leq 0$ the problem

$$A^{k+\zeta} u + \sum_1^{|k|} \psi_r YF^{-1}(P_r(\eta)) = Yf$$

gives an isomorphism $YL^2 \times C^{|k|} \to YL^2$ if and only if $\{Y^F P_r\}$ span $\mathrm{Coker}(A^{k+\zeta})\hat{}$ whose dimension is $|k|$. (succinctly, a problem for $A^{k+\zeta}$ is well posed $\Longleftrightarrow$ the covering condition is satisfied).

2. Vectorial one-dimensional problems.

We now consider the diagonal (uncoupled) system of operators

$$A^k u = (A^{k_1} u_1, \ldots, A^{k_N} u_N)$$

Here $u = (u_1, \ldots, u_N)$ is a column vector, $k = (k_1, \ldots, k_N)$ (integers) and A^k is a diagonal matrix of operators acting in $(YL^2)^N$. Obviously

$$\dim \operatorname{Ker} A^k = \sum_{k_j > 0} k_j = k_+$$

$$\dim \operatorname{Coker} A^k = \sum_{k_j < 0}' k_j = k_-$$

$$\operatorname{Index} A^k = k_+ - k_- = \sum k_j \quad .$$

The kernel, range, cokernel (or a subspace complementary to the range) are obtained as direct sums of the corresponding objects for each component. A novel situation is obtained when one considers for A^k coupled boundary conditions and coupled potentials. Each potential symbol $P_t(\eta)$, $1 \leq t \leq p$ is an N-vector (column) $(P_{1t}(\eta), \ldots, P_{Nt}(\eta))$ and the equations we pose are

$$(2.1) \qquad A^{k_j} u_j + Y \Sigma \Psi_t F^{-1} P_{jt} = Y f_j \quad , \quad 1 \leq j \leq N \ .$$

Each boundary operator has symbol B_i, $1 \leq i \leq b$, given by an N-vector (row) $(B_{i1}(\eta), \ldots, B_{iN}(\eta))$ and the boundary conditions (transformed) are:

$$(2.2) \qquad \langle Y^F \overline{B}_i, (Yu)\hat{} \rangle = \sum_{s=1}^{N} \langle Y^F \overline{B}_{is}, (Yu_s)\hat{} \rangle = \frac{\Phi_i}{2\pi} \ , \quad 1 \leq i \leq b \ .$$

Of course $B_{is}(\eta)$ and $P_{jt}(\eta)$ are L^2-functions.

If $b = k_+$ and $p = k_-$, the problem (2.1)-(2.2) determines an isomorphism $(YL^2)^N \times C^{k_-} \to (YL^2)^N \times C^{k_+}$ if and only if $Y^F\overline{B}_i(\eta)$ span $\mathrm{Ker}\,(A^k)^\wedge$ and $Y^F P_t(\eta)$ span a complementary subspace to $\mathrm{Range}(A^k)^\wedge$ ("span the cokernel"). We argue as in Lemma 1.1, 1.2: There is a (unique) projection of arbitrary) YF on Range A^k, obtained by subtracting a combination of the potentials, if and only if these potentials span a subspace complementary to the range. Being in the range we can find a solution Yu. The boundary-conditions now determine the projection of $(Yu)^\wedge$ on the space spanned by $Y^F\overline{B}_i(\eta)$. This makes the solution unique only if that space is $\ker(A^k)^\wedge$.

More interesting is the case of redundancy b k_+, p k_-. We have

THEOREM 2. The problem (2.1)-(2.2) determines an isomorphism

(2.3) $$(YL^2)^N \times C^p \to (YL^2)^N \times C^b$$

if and only if the following conditions are satisfied:

(i) $\{Y^F\overline{B}_i\}$ is linearly independent; $\{Y^F P_t\}$ is linearly independent.

(ii) $[Y^F\overline{B}_i] = \hat{B}' \oplus \hat{B}''$ where $\hat{B}' = \mathrm{Ker}(A^k)^\wedge$;

(iii) $[Y^F P_t] = \hat{P}' \oplus \hat{P}''$ where $\hat{P}'$ is complementary to $\mathrm{Range}(A^k)^\wedge$ and $\hat{P}'' = [Y^F P_t] \cap \mathrm{Range}(A^k)^\wedge$; ([ ] denotes the span of the vectors inside.)

(iv) $A^k B''$ and P'' (or their hats) are r-dimensional spaces ($r = b - k_+ = p - k_-$) and they are dual in the sense that no vector $\neq 0$ in one space is orthogonal) to the other space.

PROOF. The argument that $[Y^F \bar{B}_i]$ should contain $\mathrm{Ker}(A^k)^\wedge$ and $[Y^F P_t]$ should contain subspace was given above. So we can write down the direct sum decompositions (ii), (iii). Furthermore, if say $Y^F \bar{B}_1 = \sum_{i>1} \alpha_i Y^F \bar{B}_i$ we cannot assign arbitrary values to $\langle Y^F \bar{B}_1, (Yu)^\wedge \rangle$. If a non-trivial combination $\sum_t \alpha_t Y^F P_t = 0$, we can add α_t to ψ_t in any solution of (2.1)-(2.2) without changing the right-hand sides.

It rest to check condition (iv). Since we can subtract from the unknown Yu a fixed function Yw with a boundary data (i.e. a given projection on $B' \oplus B''$), it suffices to establish existence and uniqueness for (2.1)-(2.2) with $\psi_i = 0$ (i.e. Yu orthogonal to $B' \oplus B''$).

A given Yf has a unique projection $Yg' \in P'$ such that $Y(f-g') \in \mathrm{Range}\ A^k$. To this we add an arbitrary, yet undetermined, $-Yg'' \in P''$. Then $Y(f-g'-g'') \in \mathrm{Range}\ A^k$. Moreover we know already that its unique pre-image Yu which is orthogonal to $B' = \mathrm{Ker}\ A^k$ is $Yu = A^{-k}(f-g'-g'')$. Now Yg'' has to be determined, and uniquely, by the additional boundary conditions $\langle B'', Yu \rangle = 0$ (scalar product in $(YL^2)N$). This can be written as

$$\langle b'', A^{-k}g''\rangle = \langle b'', A^{-k}(f-g')\rangle, \quad \underline{\text{for each}} \quad b'' \in B''$$

or as A^k is the adjoint of A^{-k}

(2.4) $\langle A^k b'', Yg''\rangle = \langle b'', A^{-k}(f-g')\rangle$, <u>each</u> $b'' \in B''$.

By definition, $B'' \cap \mathrm{Ker}\, A^k = 0$. So dim $B'' = \dim(A^k B'') = r$ say. By elementary linear algebra an element $Yg'' \in P''$ is uniquely determined by its projection on $A^k B''$ (this is (2.4)), if and only if P'' is also r-dimensional and dual to $A^k B''$ (i.e. no Yg'' is orthogonal to all $A^k B''$). The theorem is now proved.

Now we can condider the general case of <u>coupled</u> 1-dimensional system with symbol

(2.5) $M(\eta) = I + K(\eta)$, $K(\eta) \in FL^1(R)$.

Such a $N \times N$ matrix-function has, by a theorem of Gohberg-Krein [3, 4, 7], decompositions

(2.6) $M(\eta) = Q_-^{-1}(\eta)\left(\frac{\eta+i}{\eta-i}\right)^k Q_+(\eta)$,

where $Q_{\pm}(\eta)$ and $Q_{\pm}^{-1}(\eta)$ are bounded have bounded holomorphic extensions to $C_{\pm}$. The middle factor is a diagonal factor - the symbol of A^k - which is uniquely determined by $M(\eta)$. We denote by $Q_{\pm}$ the operators

$$Q_{\pm} = F^{-1} Q_{\pm}(\eta) F .$$

Lemma 2.1. The operator Q_+ gives an isomorphism of Ker M ($M = YF^{-1}M(\eta)FY$) onto Ker A^k. The operator Q_- is an isomorphism of Range M onto Coker A^k.

PROOF. $Q_{\pm}$ (and their inverses) are isomorphisms of $(Y_{\pm} L^2)^N$. (Recall that $Y=Y_+$). Formally this means that $Q_{\pm}$ commutes with $Y_{\pm}$. We can write then $M=Q_-^{-1} A^k Q_+$ and clearly $Y_+ u \in \mathrm{Ker}\, M \longleftrightarrow Y_+ \bar{Q}_+ u = Q_+ Y_+ u \in \mathrm{Ker}\, A^k$. Also if $Y_+ v = M Y_+ u$ then

$$v = F^{-1} M(\eta) F Y_+ u = Q_-^{-1} F^{-1} \left(\frac{\eta+i}{\eta-i}\right)^k F Q_+ Y_+ u \ \underline{\bmod}\ (Y_- L^2)^N.$$

We operate by Q_-. Since it preserves $\bmod (Y_- L^2)^N$

$$Q_- v = F^{-1}\left(\frac{\eta+i}{\eta-i}\right)^k F(Q_+ Y_+ u) \ \underline{\bmod}\ (Y_- L^2)^N. \tag{2.7}$$

Again since $Q_- v = Q_- Y_+ v \bmod (Y_- L^2)^N$, (2.7) means that $Q_- Y_+ v = A^k (Q_+ Y_+ u)$ and so belongs to Range A^k.

As before, let $B = (B_{is}(\eta))$ be a $b \times N$ matrix of boundary-conditions-symbols and $P = (P_{tj}(\eta))$ an $N \times p$ matrix of potential-symbols. It follows readily from Lemma 2.1 and the proof of Theorem 2 that solving the problem associated with (M, B, P) is equivalent to solving the problem with A^k, BQ_+, $Q_- P$ (BQ_+ and $Q_- P$ are again in L^2). Moreover B and P satisfy all the conditions of Theorem 2 with respect to KerM and Range M if and only if BQ_+ and $Q_- P$ satisfy these conditions with respect to Ker A^k, Range A^k. We have:

THEOREM 2 bis. If $M(\eta)$ is represented by (2.6), or even by (2.6) with k replaced by $k+\zeta$ and $-1/2 < \mathrm{Re}\,\zeta < 1/2$, then Theorem 2 remains true when we substitute M for A^k throughout.

The statement concerning $k+\zeta$ follows from Lemma 1.3, as argued in Section 1. The fact is that any invertible matrix $M(\eta)$ with some smoothness conditions and having limits $\lim_{\eta \to \pm\infty} M(\eta) = M_{\pm}$

has a representation like (2.6) with $k+\zeta$ instead of k and $-1/2 < \mathrm{Re}\,\zeta_j \leq 1/2$ (cf. [7]). In fact ζ_j are $(2\pi i)^{-1}\log \lambda_j$ where λ_j are the eigenvalues of the matrix $M_+^{-1}M_-$. The requirement $\mathrm{Re}\,\zeta_j \neq 1/2$ is then that those λ_j are not real negative.

3. n-dimensional problems.

We now consider homogeneous convolution problems in the half-space $R_+^n = \{(x,y),\ y>0\}$ (Y its characteristic function). These problems are given by three homogeneous matrices, $M(\xi,\eta)$, $B(\xi,\eta)$, $P(\xi,\eta)$, which satisfy Lipschitz condition on $|\xi|^2+\eta^2 = 1$. M, B, P are $N\times N$, $b\times N$, $N\times p$ matrices, respectively. A function f is homogeneous of order s if $f(\lambda\xi,\lambda\eta) = \lambda^s f(\xi,\eta)$, $\lambda > 0$. The triplet (M,B,P) we call <u>0-homogeneous</u> if its components are homogeneous of orders $0, -1/2, -1/2$ respectively. We assume that $M(\xi,\eta)$ is invertible and moreover that

(3.1)
$$B_{is}(\xi,\eta) = B'_{is}(\xi)B''_{is}(\xi,\eta);\quad P_{jt}(\xi,\eta) = P'_{jt}(\xi)P''_{jt}(\xi,\eta),$$
B''_{is} <u>and</u> P''_{jt} <u>are homogeneous of orders</u> $< -1/2$.

This implies that for $1 \leq j,\ s \leq N,\ 1 \leq i \leq b,\ 1 \leq t \leq p$:

(3.2) $B_{is}(\overline{\xi},\eta)$ <u>and</u> $P_{jt}(\overline{\xi},\eta) \in L^2(R^1)$ for each $\overline{\xi}$, $|\overline{\xi}| = 1$.

Also $\lim_{\eta \to \pm\infty} M(\xi, \eta) = M(0, \pm 1)$ (independent of ξ).

Thus, by what was said at the end of last section, theorem 2bis can be applied to each of the 1-dimensional problems with symbols $(M, B, P)(\bar{\xi}, \eta)$ and the condition "$\zeta_j \neq 1/2$" is satisfied if the eigenvalues of $M^{-1}(0, 1)\, M(0, -1)$ are not real negative.

By successive integration one verifies that if $\psi_t(\xi) \in L^2(R^{n-1})$ then $\psi_t(\xi) P_{jt}(\xi, \eta) \in L^2(R^n)$ and if $u_s(\xi, \eta) \in L^2(R^n)$ then $\int B_{is}(\xi, \eta) u_s(\xi, \eta) d\eta \in L^2(R^{n-1})$. We consider the system

$$(3.3) \qquad YF^{-1}\Big[\sum_s M_{js}(\xi, \eta)(Yu_s)^\wedge(\xi, \eta) + \sum_t \psi_t^\wedge(\xi) P_{jt}(\xi, \eta)\Big] = Yf_j, \quad 1 \le j \le N,$$

$$(3.4) \qquad \int \sum_s B_{is}(\xi, \eta)(Yu_s)^\wedge(\xi, \eta) d\eta = \sum_s \langle \bar{B}_{is}, (Yu_s)^\wedge \rangle = \frac{\phi_i^\wedge(\xi)}{2}, \quad 1 \le i \le b.$$

The map $(Yu, \psi) \to (Yf, \phi)$ is a well defined bounded map, denoted by $\langle M, B, P \rangle$, of

$$(3.5) \qquad H_1 = YL^2(R^n)^N \times L^2(R^{n-1})^p \longrightarrow YL^2(R^n)^N \times L^2(R^{n-1})^b = H_2.$$

For each fixed $\xi \in S^{n-2}$ (the unit sphere $|\bar{\xi}| = 1$) we define the symbol triplet

$$(3.6) \qquad (M, B, P)_{\bar{\xi}}(\xi, \eta) = (M, B, P)(\bar{\xi}|\xi|, \eta),$$

the problem and the map $H_1 \to H_2$ induced by this symbol triplet is denoted by $\langle M, B, P \rangle_{\bar{\xi}}$.

LEMMA 3.1. The a-priori estimate

(3.7) $$\|h\|_{H_1} \leq K \, \| \langle M, B, P \rangle h \|_{H_2}, \quad \text{for all } h \in H_1$$

is equivalent to the family (parametrized by $\bar{\xi} \in S^{n-2}$) of a-priori estimates

(3.8) $$\|h\|_{H_1} \leq K \, \| \langle M, B, P \rangle_{\bar{\xi}} h \|_{H_2} \quad \text{for all } h \in H_1 .$$

Equivalence also holds between the a-priori estimates for the adjoint operators acting from H_2^* to H_1^* (adjoint spaces). Thus $\langle M, B, P \rangle$ is an isomorphism if and only if $\langle M, B, P \rangle_{\bar{\xi}}$ is an isomorphism for all $\bar{\xi} \in S^{n-2}$.

This Lemma is fundamental, and constitutes a natural extension of Lemma 1.1 in [7]. As we did there, we present here a proof which carries over to L^p estimates (but for this case we have to increase smoothness conditions and modify homogeneity orders and boundary spaces).

PROOF OF LEMMA 3.1. It is sufficient to prove the following: Around any $\bar{\xi} \in S^{n-2}$ there is a ball U (in the metric of S^{n-2}, say) such that

(3.9) (3.7) and (3.8) are equivalent for all $h' = \beta^F h \in H_1$ where $\beta^F = F^{-1} \beta(\xi) F$, $\beta(\xi)$ smooth and supported in U.

(It is understood that $\beta(\xi)$ is defined on S^{n-2} and extended as homogeneous of order 0 to R^{n-1}.) Indeed a <u>finite</u> number of balls U_j cover S^{n-2} and there is a smooth partition of unity $1 = \sum \beta_j(\xi)$

with Support $\beta_j \subset U_j$. Now β_j^F commutes with all the operators employed in the estimates and for each h of H_1 or H_2, $\|h\| \sim \sum_j \|\beta_j^F h\|$. (Since $1 = \sum_j \beta_j$ and β_j^F is bounded in L^p, i.e. β_j is a multiplier in FL^p, $1 < p < \infty$). We call the estimate for all $\beta^F h$ a $\bar{\xi}$-restricted estimate and observe that for $\langle M, B, P\rangle_{\bar{\xi}}$, a ξ'-restricted estimate implies ξ''-restricted estimate since the symbols depend only on $|\xi|$.

We prove (3. 9) first for dimension n=2. Then $S^{n-2} = \{\pm 1\}$. We take the partition of unity as $e_+(\xi)$ and $e_-(\xi)$, the characteristic functions of $\xi > 0$ and $\xi < 0$ respectively. Both are multipliers in FL^p. For $e_+^F h$ resp. for $e_-^F h$, the operators $\langle M, B, P\rangle$ and $\langle M, B, P\rangle_{+1}$ [resp. $\langle M, B, P\rangle_{-1}$] actually coincide.

If $h > 2$, the two operators we compare (i.e. their symbols) are not the same but sufficiently close on U if U is small. To be precise, we argue as follows: We may assume $\xi = (1, 0, \ldots, 0)$. We construct a family of "cut-functions"

$$\Gamma_\delta(\xi) = \Gamma(\xi_1, \xi_2/\delta, \ldots, \xi_{n-1}/\delta)$$

where $\Gamma = \Gamma_1$ is a fixed smooth function on S^{n-2} (extended in a homogeneous fashion), which is 1 for $|\xi - \bar{\xi}| < 1/2$ and 0 for $|\xi - \bar{\xi}| > 1$. The corresponding domains for Γ_δ are still neighbourhoods of $\bar{\xi}$ which shrink to $\bar{\xi}$ as $\delta \to 0$. We shall prove that

$$\text{(3. 10)} \qquad \lim_{\delta \to 0} \| \langle M, B, P\rangle \Gamma_\delta^F - \langle M, B, P\rangle_{\bar{\xi}} \Gamma_\delta^F \| = 0$$

(in the operator norm). Then for small δ, the norm of the difference of the two operators is small compared to the norm of each operator and the (unrestricted) a-priori estimates for one implies the estimate for the other. For such Γ, we can choose U inside the domain where $\Gamma_\delta(\xi) \equiv 1$, then $\Gamma_\delta(\xi)\beta(\xi) = \beta(\xi)$ if Support $\beta \subset$ U and the U-restricted estimates for $\langle M, B, P\rangle$ and $\langle M, P, B\rangle_{\xi}$ become equivalent.

To prove (3. 10) it suffices to show that the symbols

$$(3.11)\quad [(M,B,P)(\xi_1,\dots,\xi_{n-1},\eta)-(M,B,P)(|\xi|,0,\dots,0,\eta)]\Gamma(\xi_1,\xi_2/\delta,\dots,\xi_{n-1}/\delta)$$

converge to 0 in the multiplier norm (in FL^p). This norm, being invariant under stretching of each variable, is equal to the norm of

$$[(M,B,P)(\xi_1,\delta\xi_2,\dots,\delta\xi_{n-1},\eta)-(M,B,P)(|\xi|,0,\dots,0,\eta)]\Gamma(\xi_1,\dots,\xi_{n-1}).$$

Here the norm of the bracketed term tends to 0 due to the smoothness of (M, B, P), while the norm of Γ is fixed.

This concludes the proof of the first equivalence. We can argue in a quite abstract fashion that the same proof works for the equivalence of the adjoint estimates. We have essentially used the facts that

(i) β_i^F are bounded and commute with all the estimated operators;

(ii) stretching of ξ-variables preserve the (multiplier) norm;

(iii) a certain family (3. 11) converge to 0 in the operator norm. All these facts are preserved under taking adjoints. Actually the adjoints of β_i^F and stretching are operators of the same type.

Now a problem $\langle M, B, P\rangle_{\bar{\xi}}$, although n-dimensional, has symbols depending on two real variables, $|\xi|$ and η, only. Due to homogeneity, on the unit sphere they depend on one variable, and it is natural to compare that problem with the one-dimensional one with symbol $(M, B, P)(\bar{\xi}, \eta)$. A straighforward application of Parseval identity (with respect to x and ξ) shows that indeed one of them gives an isomorphism if and only if the other does. (We omit this easy proof since a more general one exists for the L^p-theory, following the ideas of [7] and to presented in [8]). Using Theorem 2bis and Lemma 3.1 we arrive at

THEOREM 3. $\langle M, B, P\rangle$ is an isomorphism of H_1 onto H_2 if and only if the eigenvalues of $M^{-1}(0,1)M(0,-1)$ are not real negative and for each $\bar{\xi} \in S^{n-2}$ the 1-dimensional symbols $(M, B, P)(\bar{\xi}, \eta)$ satisfy the "covering conditions" (i)-(iv) of Theorem 2 (with M replacing A^k).

4. Various remarks.

We have chosen (M, B, P) as 0-homogeneous, in particular B and P of order -1/2, in order to get pure L^2 estimates. This is not essential. Keeping first M of order 0, we can take B_i (each B_{is}) of order β_i and P_t (each P_{jt}) of order α_t, provided (3.1) is satisfied. For multiplying B_i by $|\xi|^{-1/2-\beta_i}$ we return to the previous case. This involves a change in the domain H_1 and the range H_2 (cf. (3.5)) of the isomorphism. The domain becomes $L^2(R^n)^N \times \prod_t H^{\alpha_t+1/2}(R^{n-1})$ and the range becomes

$L^2(R^n)^N \times \prod_i \overset{\circ}{H}^{\beta_i+1/2}(R^{n-1})$ where $\overset{\circ}{H}^s$ is the closure of the C_o^∞-functions under the norm $[\int |\Phi^\wedge(\xi)|^2 |\xi|^{2s} d\xi]^{1/2}$ (For $s>0$ this is also H^s modulo polynomials of degree $>s$.) The norms H^s and $\overset{\circ}{H}^s$ are equivalent for functions supported in a fixed bounded domain. Thus one can use $\overset{\circ}{H}^s$-estimate as well (and they are more convenient) to obtain general estimates in a bounded domain.

Next we treat certain systems (M, B, P) in which the orders of homogeneity of B_{is}, P_{jt} and M_{js}, $1 \leq j, s \leq N$ depend also on j and s like in Douglis-Nirenberg elliptic systems (cf. [1, 5]). One uses real weights γ_j and δ_s and requires that order $M_{js} = \gamma_j + \delta_s$, order $B_{is} = \delta_s + \beta_i$, order $P_{jt} = \gamma_j + \alpha_t$. If we make the substitutions

$$(Yu_s) = (\eta - i|\xi|)^{-\gamma_s}(Yv_s)^\wedge, \qquad (Yf_j)^\wedge = (\eta - i|\xi|)^{\delta_j}(Yg_j)^\wedge$$

in a system of the form (3. 3)-(3. 4) and in addition $\psi_t^\wedge = |\xi|^{-\alpha_t - 1/2} \tilde{\psi}_t^\wedge$, $\Phi_i^\wedge = |\xi|^{\beta_j + 1/2} \tilde{\Psi}_i^\wedge$ (this is the previously described modification) then $(v, \tilde{\psi}, g, \tilde{\Phi})$ satisfy a 0-homogeneous system. The isomorphism is now between

$$\prod_s Y\overset{\circ}{H}^{\gamma_s}(R^n) \times \prod_t \overset{\circ}{H}^{\alpha_t}(R^{n-1}) \rightarrow \prod_j Y\overset{\circ}{H}^{\delta_j}(R^n) \times \prod_i \overset{\circ}{H}^{\beta_i}(R^{n-1}).$$

Thus our systems include in particular most general differential elliptic problems and the covering condition we have found generalizes the covering condition for that case (cf. the formulation in [1, 2, 5]). In that case one takes only boundary conditions because (for each ξ) the indices ζ_j are positive. (In particular we have outlined a proof of the L^p estimates for those problems). Some readers may wonder why the behaviour of $\overline{B}_i(\eta)$ in C_+ i.e. of $B_i(\eta)$ in C_- (contrary to [1, 2]) enters in the covering condition. This is because we chose Fourier transform with + .

For a given system, say 0-homogeneous, we now ask about solvability in H^s instead of $H^o = L^2$. However, unless the operator M preserves supports (in R^n_-) there is a difficulty is defining the operator in $YH^s = H^s(\bmod Y_- H^s)$, different extensions of Yu to $u \in H^s$ may lead to different results. (Only for $|s| < 1/2$ we can take the extension as zero in R^n_-).

As in [7], we can overcome this difficulty by considering the operator $\tilde{M} : u \to (Y_- u, Y_+ Mu)$ mapping $H^s \longrightarrow Y_- H^s \times Y_+ H^s$ (the potentials $Y_t P \psi$ are added to the 2nd component).

As shown in [7] , problem for $\tilde{M}$ in H^s is equivalent to an L^2- problem for the operator with symbol $M_s = (\eta - i|\xi|)^{-s} M(\eta + i|\xi|)^s$. If M had $A^{k+\zeta}$ in its factorization then M_s has $A^{k+\zeta -s}$. We observe now that in case (M, B, P) failed to be an L^2-isomorphism only because $\operatorname{Re} \zeta_j = 1/2$ for some j , it will be an isomorphism when we pass to H^s, s near 0 (in fact for all s in the range $|s| < 1/2$ but for at most N exceptional values). If more boldly we increase s by integral steps we ultimately get all indices $\zeta_j - s$ of M_s non-positive. The $\tilde{M}$ will have a trivial kernel in H^s and we can get well-posed problems with potentials P_t only. (Dually for negative s we ultimately get trivial cokernel in H^s and problems with boundary conditions B_j only). When s further increases to s+1 , the number ρ of needed potentials increase by N , i.e. the dimension of the cokernel of the associated 1-dimensional problems increase by N. This actually implies the non-existence of "regularization" for the solution of our convolution problems. Formally this results from the fact that the range space $Y_{-}H^{s} \times Y_{+}H^{s}$ is "too big", but for operators which do not preserve support there does not seem another reasonable choice.

Finally we remark that if (M, B, P) satisfies the covering condition (for each $\bar{\xi} \in S^{n-2}$) then Index $M(\bar{\xi}, \eta) =$ $=$ Ker $M(\bar{\xi}, \eta)$-Coker $M(\bar{\xi}, \eta)$ is a trivial class of the Grothendick group $K(S^{n-2})$. In fact it is equal to the difference of the two trivial vector bundles $[Y^{F}\bar{B}_{j}(\xi, \eta) - Y^{F} P_{t}(\bar{\xi}, \eta)]$. This follows easily from condition (i)-(iv) of Theorem 2. The converse is also true. If Index $M(\bar{\xi}, \eta)$ is a trivial class, appropriate B_j and P_t can be found. This follows again from Theorem 2 and the fact two r-dimensional spaces are "in general" dual. A similar triviality condition is also the criterion for existence of well-posed problems for general elliptic pseudo-differential systems in a bounded domain. This was also proved in [10].

E. Shamir

BIBLIOGRAPHY

1. S. Agmon, A. Douglis and L. Nirenberg, Estimates near the boundary for solutions of elliptic partial differential equations satisfying general boundary conditions II, Comm. Pure Appl. Math. 17 (1964), 35-92.

2. M. F. Atiyah and R. Bott, The index problem for manifolds with boundary, Bombay Coll. Diff. Analysis, Oxford Univ. Press (1964), 175-186.

3. I. C. Gohberg and M. G. Krein, Systems of singular integral equations on the half line with kernels depending on the difference of the arguments, Usp. Mat. Nauk 13 (1958), 3-72.

4. I. C. Gohberg, Factorization problems in normed rings, Usp. Mat. Nauk 19 (1964), 71-124.

5. L. Hörmander, Linear Partial Differential Operators, Academic Press and Springer, 1963.

6. E. Shamir, Une proprieté des espaces $H^{s,p}$, C. R. Acad. Sci., Paris 255 (1962), 448-449.

7. ________, Elliptic systems of singular integral operators. I. The half-space case, Tran. Amer. Math. Soc. 127 (1967), 107-124.

8. ________, Elliptic systems of singular integral operators . II. Boundary value problems in a half-space . (to appear)

9. M. I. Visik and G. I. Eskin, Convolution equations and systems of equations . Usp. Mat. Nauk 22 (1967), 15-76.

10. ________, Normally solvable problems for elliptic convolution systems. Mat. Sbornik 74 (116), (1967), 326-356.

CENTRO INTERNAZIONALE MATEMATICO ESTIVO

(C. I. M. E.)

I. M. SINGER

ELLIPTIC OPERATORS ON MANIFOLDS

Corso tenuto a Stresa dal 26 Agosto al 3 Settembre 1968

Elliptic Operators on Manifolds

by I. M. Singer

1. Introduction. In these talks I shall give examples of some applications of elliptic operators on manifolds centering around the fixed point formula and the index theorem. In his lectures at this session, R. T. Seeley will review pseudo-differential operators on manifolds and the elementary properties of the index. This will allow me to assume much of the analytical facts concerning elliptic operators. Since the primary discipline of the majority of the participants is analysis, I will spend more time with the geometry and topology but motivate the proofs of the main theorems from the analytic viewpoint. By and large analysts find the geometrical and topological background needed here formidable. It is my hope that these expository talks will help orient the listener and allow him to read the relevant papers listed at the end more easily.

2. Examples. Elliptic operators have many applications to manifold theory because much of the geometric and topological structure of a manifold is reflected by certain elliptic systems on the manifold. There are only a few critical examples. Let us review them before discussing the general case. For simplicity, we assume our manifolds are oriented.

Example 1. The prototype of an elliptic complex is the De Rham complex. Here the differential operator is the exterior derivative d which, in local coordinates in R^n, operates or a p-form

$$\sum_{i_1,\ldots,i_p=1}^{n} a_{i_1\ldots i_p} \, dx_{i_1} \wedge\ldots\wedge dx_{i_p}$$ to give a $p+1$ form

$$\sum_{i_1,\ldots,i_p,j=1}^{n} \frac{\partial a_{i_1\ldots i_p}}{\partial x_j} dx_j \wedge dx_{i_1} \wedge\ldots\wedge dx_{i_p} .$$ If X is a C^∞-manifold, the space of p-forms at each point forms a family of vector spaces $\Lambda^p = \{\Lambda^p_x\}_{x\in X}$ parametrized by X, which in any coordinate neighborhood U looks like $U \times R^{\binom{n}{p}}$ with the transition functions on $U \cap V$ given by C^∞ matrix functions on $U \cap V$. A smooth p-form is a smooth section of Λ^p and the space of all such sections will be denoted by $C^\infty(\Lambda^p)$. On X, the total differential becomes a first order differential operator $d\colon C^\infty(\Lambda^p) \to C^\infty(\Lambda^{p+1})$. If we put these maps together

for all p, $0 \leq p \leq \dim X = n$, we get the De Rham complex:

$$0 \to C^\infty(\Lambda^0) \xrightarrow{d} C^\infty(\Lambda^1) \xrightarrow{d} \cdots \xrightarrow{d} C^\infty(\Lambda^p) \xrightarrow{d} C^\infty(\Lambda^{p+1}) \xrightarrow{d} \cdots \xrightarrow{d} C^\infty(\Lambda^n) \xrightarrow{d} 0 .$$

It enjoys these properties:

(i) $d^2 = 0$.

(ii) locally the Poincare lemma holds; i.e., if $\omega \in C^\infty(\Lambda^p)$ and $d\omega = 0$ in some neighborhood U of x, then there exists a neighborhood $V \subset U$ of x and a $u \in C^\infty(\Lambda^{p-1}) \ni du = \omega$ on V.

(iii) The symbol sequence:

$$0 \to \Lambda^0 \xrightarrow{\sigma_d(\xi)} \Lambda^1 \xrightarrow{\sigma_d(\xi)} \Lambda^2 \to \cdots \cdots \xrightarrow{\sigma_d(\xi)} \Lambda^n \to 0 \quad \text{is exact} .$$

What we mean is this: Let $\sigma_d(\xi)$ denote the characteristic polynomial or symbol of d on the cotangent vector ξ. (For a first order operator D, it is easy to check that when $\xi = du$ at the point x and $u(x) = 0$, then $(\sigma_D(\xi))(f) = i(D(uf))(x)$.) Exactness means $\ker \sigma_d(\xi): \Lambda^p \to \Lambda^{p+1}$ is equal to the image of $\sigma_d(\xi): \Lambda^{p-1} \to \Lambda^p$ at each point of $x \in X$, $0 \leq p \leq n$, and $\xi \neq 0$. In this example, we do get exactness because $i(d(u\omega)) = i(du\,\omega + ud\omega)$ so that $\sigma_d(\xi) = i \times$ left multiplication by $\xi = i\, \iota_\xi$. The well known algebraic property of Grassman algebras says that $\ker \iota_\xi: \Lambda^p \to \Lambda^{p+1}$ equals image $\iota_\xi: \Lambda^{p-1} \to \Lambda^p$.

(iv) Since $d^2 = 0$, ker d: $C^\infty(\Lambda^p) \to C^\infty(\Lambda^{p+1})$ contains image of d: $C^\infty(\Lambda^{p-1}) \to C^\infty(\Lambda^p)$. Let $H^p(M) = \ker d/\text{image } d$. Assume M is compact. The celebrated De Rham theorem says $H^p(M)$ is naturally isomorphic with the p-th cohomology of M with real coefficients. In particular, dim $H^p(M) = b_p$ the pth Betti number of M and the Euler characteristic $\chi(m)$ equals $\sum (-1)^p \dim H^p(M)$.

(v) Suppose f: $M \to M$ is a C^∞ map. Then f induces a map df_p from $\Lambda^p_{f(x)} \to \Lambda^p_x$ where in local coordinates

$$df_p(\sum a_{i_1..i_p} dx_{i_1} \wedge .. dx_{i_p}) = \sum a_{i_1..i_p} \circ f \;\; d(x_{i_1} \circ f)\,.$$

This map commutes with the total differential d and hence induces a linear transformation $\tilde{f}_p: H^p(M) \to H^p(M)$. Let the Lefschetz number L(f) be $\sum (-1)^p \operatorname{tr}(\tilde{f}_p)$. The classical Lefschetz fixed point formula says that if f has simple fixed points (i.e. at each fixed point x, the map $df_1: \Lambda^1_x \to \Lambda^1_x$ does not have 1 in its spectrum; equivalently, the graph of f intersects the diagonal of X x X **transversally**.), then

$$L(f) = \sum_{\text{fixed points}} \pm 1$$

the sign depending on the sign of the determinant of $1-df_1$ at the fixed point x. In particular, if $L(f) \neq 0$, then f has a fixed point.

Remark. The family of vector spaces Λ^p is the prototype for a real vector bundle over X , which we now define. A <u>vector bundle</u> W over X is a topological space whose point set is $\bigcup_{x \in X} W_x$ where $\{W_x\}_{x\in X}$ is a family of k-dimensional vector spaces over X . The family must fit together as follows: Let $\pi: W \to X$ which sends W_x into x . There exists a covering U and maps $\{\phi_u\}_{u\in U}$ where

$$\begin{array}{ccc} \pi^{-1}(u) & \xrightarrow{\phi_u} & u \times R^k \\ \pi \downarrow & & \downarrow \\ u & \xrightarrow{I} & u \end{array}$$

is commutative.

The maps $\{\phi_u\}_{u\in U}$ fit together for $u,v \in U$. The transition functions $\phi_v \circ \phi_u^{-1} : v \cap u \times R^k \to v \cap u \times R^k$ is a linear transformation in the second variable so that $\phi_v \circ \phi_u^{-1} (x \times \overline{r}) = (x, a_{v,u}(x)(\overline{r}))$ where $a_{v,u}: v \cap u \quad k \times K$ real matrices is continuous. Complex vector bundles are the same as above with the real field replaced by the complexes. When X is a a (complex), manifold, W is a C^∞ (holomorphic) vector bundle if $a_{v,u}$ is C^∞ (holomorphic) .

For example, for Λ^1 we can take U to be coordinate patches. If u is a coordinate patch with coordinates $x_1,\dots,x_n$, then $\phi_u: \pi^{-1}(u) \to U \times R^k$ maps $\omega = \sum a_i dx_i \to (x, a_1, \dots, a_n)$

For another coordinate patch V with coordinates $y_1,\dots,y_n$ and $x_i = \Psi_i(y_j)$ on $u \cap v$ we have $dx_i = \sum \frac{\partial \Psi_i}{\partial y_j} dy_j$ so that $\omega = \sum a_i dx_i = \sum a_i \frac{\partial \Psi_i}{\partial y_j} dy_j$. Hence in this case $a_{v,u} = \frac{\partial \Psi_i}{\partial y_j}$.

The space $C^\infty(W)$ is the space of maps $s: X \to W$ such that $\pi \circ s$ equals the identity map on X and s is locally C^∞. Analytically this means for each $u \in \mathcal{U}$, s is given by a k-tuple of C^∞ maps s_u (where $\Phi_u \circ s(x) = (x, s_u(x)), x \in U$) and on overlaps $u \cap v$, $a_{v,u}\, s_u = s_v$.

Example 2. Frequently, if the manifold has additional structure, other elliptic complexes can be constructed. The classical example is a complex n-manifold where one has coordinates $z_1,\dots z_n, \bar{z}_1,\dots,\bar{z}_n$. Then locally $\Lambda^p \otimes C$ (p-forms with complex coefficients) splits into $\sum_{r \le p} \oplus \Lambda^{r,p-r}$ where $\Lambda^{r,p-r}$ denotes those forms which are linear combinations of the type $dz_{j_1} \wedge \dots \wedge dz_{j_r} \wedge d\bar{z}_{k_1} \wedge \dots \wedge d\bar{z}_{k_{p-r}}$. Because coordinate changes are holomorphic, this decomposition persists globally so that the complex vector bundles $\Lambda^{p,q}$, $0 \le p,q \le n$ are well defined. The exterior differential d equals $\partial + \bar{\partial}$ where the holomorphic pact ∂ map $C^\infty(\Lambda^{p,q}) \to C^\infty(\Lambda^{p+1,q})$ and the antiholomorphic part $\bar\partial$ maps $C^\infty(\Lambda^{p,q}) \to C^\infty(\Lambda^{p,q+1})$. In local coordinates, for example, $\bar\partial(\sum a_{j_1..j_r,k_1..k_{p-r}} dz_{j_1}\wedge..\wedge dz_{j_r}\wedge d\bar z_{k_1}\wedge..\wedge d\bar z_{k_{p-r}}) =$

$$\sum \frac{\partial a_{j_1..j_r,k_1..k_{p-r}}}{\partial \bar z_l} d\bar z_l \wedge dz_{j_1}\wedge..\wedge dz_{j_r} \wedge d\bar z_{k_1}\wedge..\wedge d\bar z_{k_{p-r}} .$$

In this way, we obtain the $\bar\partial$ complex:

$$0 \to C^\infty(\Lambda^{0,0} = \Lambda^0) \xrightarrow{\bar\partial} C^\infty(\Lambda^{0,1}) \to C^\infty(\Lambda^{0,2}) \to .. \xrightarrow{\bar\partial} C^\infty(\Lambda^{0,n}) \to 0 .$$

It enjoys these properties

(i) $\bar\partial^2 = 0$

(ii) The local Poincare lemma holds.

(iii) The symbol sequence is exact:

$$0 \to \Lambda^{0,0} \xrightarrow{\sigma_{\bar\partial}(\xi)} \Lambda^{0,1} \xrightarrow{\sigma_{\bar\partial}(\xi)} \Lambda^{0,2} \to .. \xrightarrow{\sigma_{\bar\partial}(\xi)} \Lambda^{0,n} \to 0 \quad \text{for } \xi \neq 0 .$$

In this case, it is easy to verify that $\sigma_{\bar\partial}(\xi) = P\,\sigma_d(\xi)$ where $\sigma_d(\xi)\colon\ \Lambda^{0,k} \to \Lambda^{1,k} \oplus \Lambda^{0,k+1} \subset \Lambda^{k+1}$ and P is the projection onto $\Lambda^{0,k+1}$. From this the exactness is trivial for any real nonzero cotangent vector ξ has nonzero components in both $\Lambda^{0,1}$ and $\Lambda^{1,0}$.

(iv) $H^{0,p}(M)$ can be defined as before, and as we shall see from the Hodge theory, is finite dimensional. The <u>arithmetic genus</u> of M is $\sum_p (-1)^p \dim H^{0,p}(M)$.

(v) If $f: M \to M$ is holomorphic, then df_p preserves the finer structure $\Lambda^{r,p-r}$. Since it commutes with $\bar{\partial}$, df induces a map $\tilde{f}_p^h$: $H^{o,p}(M) \to H^{o,p}(M)$. Denote $\sum_p (-1)^p \operatorname{tr}(\tilde{f}_p^h)$ by $L(f)$. The generalized fixed point theorem

of Atiyah and Bott applied to this case will express $L(f)$ when f has simple fixed points, in terms of df_1 at the fixed points

(vi) If in addition to the above, we have a holomorphic vector bundle W over X, i.e., whose transition functions are complex analytic, then $\Lambda^{p,q} \otimes W$ are well defined and $\bar{\partial}: C^\infty(\Lambda^{p,q} \otimes W) \to C^\infty(\Lambda^{p,q+1} \otimes W)$ with $\bar{\partial}^2 = 0$. We can then define $H^{o,p}(M,W)$, the $\bar{\partial}$-cohomology with coefficients in W, of importance in algebraic geometry.

The above two examples lead to the

Definition: An elliptic complex $\mathcal{E}$ is a i sequence $\{E_i\}_{i=0}^N$ of C^∞-complex vector bundles over X together with a sequence of pseudo-differential operators $D_i: C^\infty(E_i) \to C^\infty(E_{i+1})$ such that (1): the symbol sequence: $0 \to E_0 \xrightarrow{\sigma D_0(\xi)} E_1 \xrightarrow{\sigma D_1(\xi)} \cdots \xrightarrow{\sigma D_{N-1}(\xi)} E_N \to 0$ is exact for each $\xi \neq 0$ at each $x \in X$; (2): $D_{i+1} D_i = 0$. For simplicity, we shall assume $\{D_i\}_{i=0}^N$

have the same true order.

Note that for $N = 1$, the two step sequence $C^\infty(E_0) \xrightarrow{D_0} C^\infty(E_1)$ being elliptic simply means that D_o is an elliptic operator. From condition (2), $\ker D_i \supseteq \text{image } D_{i-1}$ so that we can define the ith cohomology $H^i(\mathcal{E})$ as $\ker D_i/\text{image } D_{i-1}$. We shall see shortly that because $\mathcal{E}$ is elliptic, the spaces $H^i(\mathcal{E})$ are finite dimensional. Let $\chi(\mathcal{E}) = \sum_{i=0}^{N} (-1)^i \dim H^i(\mathcal{E})$ and note that when $N = 1$; $\chi(\mathcal{E}) = \dim \ker D_o - \dim \operatorname{cok} D_o = \operatorname{index} D_o$.

Both the De Rham and $\bar{\partial}$-complex are elliptic complexes. The adjective elliptic comes from the following device. We can construct an elliptic operator on $\sum_{i=0}^{N} E_i$ as follows: For convenience, assume M is orientable with a given orientation. Choose a Riemannean metric on M and Hermetian metrices on all E_i. Then the formal adjoint $D_i^*: C^\infty(E_{i+1}) \to C^\infty(E_i)$ is well defined. Consider the operator $D + D^* = \sum_i (D_i + D_{i-1}^*): C^\infty(E) \to C^\infty(E)$ where $E = \sum \oplus E_i$. It is elliptic, for $\sigma_{D_i+D_{i-1}^*}(\xi) = \sigma_{D_i} + \sigma_{D_{i-1}^*}(\xi) = \sigma_{D_i}(\xi) + (\sigma_{D_{i-1}}(\xi))^*$. If $\sigma_{D_i+D_{i-1}^*}(\xi)(e_i) = 0$, then e_i lies in the kernel of $\sigma_{D_i}(\xi)$ and in the kernel of $(\sigma_{D_{i-1}}(\xi))^*$. But $\sigma_{D_i}(\xi)e_i = 0$ implies, by the exactness of the symbol sequence, that $e_i = \sigma_{D_{i-1}}(\xi)e_{i-1}$ with e_{i-1} E_{i-1} at x. Since the kernel of $(\sigma_{D_{i-1}}(\xi))^*$ is the orthogonal complement of the image of $\sigma_{D_{i-1}}(\xi)$, we conclude that $\sigma_{D_{i-1}}(\xi)e_{i-1} = e_i = 0$ so that $D_i + D_{i-1}{}^*(\xi)$ has zero kernel for $\xi \neq 0$. Hence $\sigma_{D+D^*}(\xi)$ is nonsingular.

Other elliptic operators can be built out of $D + D^*$ as follows. Let $\tau: E \to E$ be an automorphism with $\tau^2 = I$ such that $\tau(D+D^*) = -\tau(D+D^*)$. Let $E_{\pm}$ be the ± 1 eigenspace of τ so that $E = E_+ \oplus E_-$. Since $D + D^*$ anticommutes with τ, $D + D^*: C^\infty(E_{\pm}) \to C^\infty(E_{\mp})$. Since $D + D^*$ is elliptic on all of E, $\sigma_{D+D^*}(\xi)$ for $\xi \neq 0$ is injective: $E_{\pm} \to E_{\mp}$. Hence $\dim E_+ = \dim E_-$ and $D + D^*: C^\infty(E_{\pm}) \to C^\infty(E_{\mp})$ is elliptic.

For example, suppose $\tau: E_i \to E_i$ equals $(-1)^i I$. Since D_i (and D^*_{i-1}) interchange even E's with odd E's, the anti-commuting holds. In this case, $E_+ = \sum_j \oplus E_{2j}$ and $E_- = \sum_j \oplus E_{2j+1}$ so that $D + D^*: C^\infty(E_+) \to C^\infty(E_-)$ is elliptic whenever one has an elliptic complex.

Example 3. A more esoteric case of such an involution τ comes from the De Rham complex when $\dim X = 2k$. Choose a metric on X from which one obtains the star operator $*_p: \Lambda^p \to \Lambda^{2k-p}$. An elementary computation shows $*_{2k-p} *_p = (-1)^p I$. Hence, if we let $\tau = \sum_p i^{p(p-1)+k} *_p$, we get $\tau^2 = I$. On the other hand $d^* = - * d *$, so that τ anticommutes with $d + d^*$ and gives a new elliptic complex $d + d^*: C^\infty(\Lambda_+) \to C^\infty(\Lambda_-)$. This elliptic operator has considerable importance in topology as we shall see.

Example 4. The final example of an elliptic complex stems from Dirac who wanted to find (in four dimensions) a differential operator which is a square root of the Laplacian Δ on systems. In Euclidean space R^k, given $\Delta = -\sum_{i=1}^{k} \frac{\partial^2}{\partial X_i}$, one wants to find $D = \sum_{i=1}^{k} A_i \frac{\partial}{\partial X_i}$ where A_i are square matrices such that $D^2 = \Delta$. This forces (i): $A_i^2 = -I$ and (ii): $A_i A_j + A_j A_i = 0$ for $i \neq j$. One defines the Clifford algebra C_k over $\mathbf{R}^k$ as the formal algebra with unit over the complexes generated by $A_1, \dots, A_K$ satisfying (i) and (ii). It is easy to show that $\dim C_k = 2^k$ and that C_k is a simple algebra if k is even and a direct sum of two simple algebras when k is odd. When k is even, then, C_k is a full matrix algebra over the complexes acting on a vector space of dim $2^{k/2}$ which can be taken as a minimal left ideal S of C_K. Then the operator $D = \sum_{d=1} A_j \frac{\partial}{\partial X_j} : C^\infty(S) \to C^\infty(S)$ gives an elliptic operator on R^k for $\sigma_D(\xi) = i\,(\Sigma A_j \xi_j)$ which has the inverse $i \frac{\Sigma A_j \xi_j}{|\xi|^2}$. Again, this operator can be refined by using the involution $(-1)^{k/2} A_1 A_2 \dots A_k$ on S, which anticommutes with D. Thus one gets $D: C^\infty(S_+) \to C^\infty(S_-)$.

To obtain this elliptic operator on a Riemanian manifold, one must have the family of vector spaces S_x associated with each tangent space T_x fitting together properly to form a vector bundle. This imposes an additional structure on M, called a spin structure. There is an obstruction to doing this (the 2nd Stiefel-Whitney Class) and when one can, it is possible to do it in different ways. At any rate, if M has a given spin structure, then one can construct an elliptic operator, the Dirac operator

$D : C^\infty(S) \to C^\infty(S)$ and when $\dim(M)$ is even refine it to give $D : C^\infty(S_+) \to C^\infty(S_-)$.

In the first two examples of an elliptic complex, we showed how a structure preserving map $f : M \to M$ induced a map f on the cohomology of the complex. Let us now define this **notion** for a general elliptic complex $\mathcal{E}$. By an endomorphism T of $\mathcal{E}$, we shall mean a sequence of C^∞ maps $T_i : C^\infty(E_i) \to C^\infty(E_i)$ such that $T_{i+1} D_i = D_i T_i$.

This commuting condition implies that T_i leaves $\ker D_i$ and image D_{i-1} invariant and hence induces a linear transformation T_i on $H^i(\mathcal{E})$. We will let $L(T)$ denote $\Sigma(-1)^i \operatorname{tr}(T_i)$. In many applications, T is an isomorphism, so that T^{-1} is also an endomorphism of $\mathcal{E}$ and T generates a cyclic group of **endomorphisms.** More generally, we can consider a group G of **endomorphisms** of $\mathcal{E}$. Then the map $T \to T_i$, $T \in G$ is a group representation of G on the vector space $H^i(\mathcal{E})$, whose character is the function $T \to \operatorname{tr}(T_i)$ Hence $T \to L(T)$ is an element of the character ring of G . When $T = I$, note that $L(T) = \Sigma(-1)^i \dim H^i(\mathcal{E})$, the Euler characteristic of $\mathcal{E}$.

Both the fixed point formula and the index theorem give formulas for $L(T)$, but under different assumptions. As we shall see, in the fixed point formula, one assumes that the map $f: M \to M$ induced by T has simple fixed points and the formula for $L(T)$ involves only the behavior of T at the fixed points but no assumptions are made about G . In the index theorem, one assumes G is compact but no assumptions are made about the fixed point behavior. The formula for the function $T \to L(T)$ involves the symbol of the operators in the complex $\mathcal{E}$. One can carry the index theorem a bit further and get a "fixed point" formula involving certain elliptic operators along the fixed submanifolds for each $T \in G$.

3. First applications of ellipticity. Let us quickly review what information the theory of elliptic pseudodifferential operators on compact manifolds tells us about elliptic complexes.

(a) If $Q: C^\infty(E_0) \to C^\infty(E_1)$ is an elliptic operator of degree s, then Q has a parametrix $P_1: C^\infty(E_1) \to C^\infty(E_0)$ which is an elliptic operator of degree $-s$ such that $PQ = I - C_1$ and $QP = I - C_2$ where C_1 and C_2 are of order $-\infty$, i.e., integral operators with C^∞ kernel.

(b) If instead of a 2-step elliptic complex as in (a), we have a general one $\mathcal{E}$ with operators $\{D_i\}_{i=0}^{N-1}$, then there exists a parametrix $\{P_i\}_{i=0}^{N-1}$ with $P_i: C^\infty(E_{i+1}) \to C^\infty(E_i)$ such that $D_{i-1}P_{i-1} + P_iD_i = I - C_i$ where C_i are of order $-\infty$. In fact, let $\Delta_i = D_i^*D_i + D_{i-1}D_{i-1}^* : C^\infty(E_i) \to C^\infty(E_i)$, which is elliptic. Let Q_i be a parametrix for Δ_i, and let $P_i = Q_iD_i^*$. Then $P_iD_i + D_{i-1}P_{i-1} = Q_iD_i^*D_i + D_{i-1}Q_{i-1}D_{i-1}^* = Q_i\Delta_i + (D_{i-1}Q_{i-1} - Q_iD_{i-1})D_{i-1}^*$ But $D_{i-1}\Delta_{i-1} = \Delta_iD_{i-1}$ so that $Q_iD_{i-1}\Delta_{i-1}Q_{i-1} = Q_i\Delta_iD_{i-1}Q_{i-1}$ and $Q_iD_{i-1} - D_{i-1}Q_{i-1}$ is of order $-\infty$. Hence $P_iD_i + D_{i-1}P_{i-1} = I - C_i$ with C_i of order $-\infty$.

We can actually choose the parametrix P as a function of t so that $C_i(t) \to I$ as $t \to 0$ boundedly in the strong operator topology on $C^\infty(E_i)$. This can be done as follows. Let $R_i(t)$ be a family of operators of order $-\infty$ so that $R_i(t) \to I$ as $t \to 0$ boundedly in the strong operator topology. (Such a family exists locally as a convolution operator and is obtained globally by a partition of unity.) Let $P_i(t) = (I - R_i(t))P_i$, which also gives a parametrix for $\mathcal{E}$ Then $I - C_i(t) = P_i(t)D_i + D_{i-1}D_{i-1}(t)$ $\to 0$ as desired, since $P_i(t) \to 0$.

(c) If P is elliptic of degree s, then P induces $P_r: H^r(E_0) \to H^{r-s}(E_1)$, a Fredholm operator and the regularity theorem tells us in particular that $\ker P_r \subset C^\infty(E_0)$ and $\operatorname{cok} P_r \subseteq C^\infty(E_1)$ so that $P: C^\infty(E_0) \to C^\infty(E_1)$ has finite dimensional kernel and cokernel.

(d) If P is elliptic, then $\operatorname{ind} P = \dim \ker P - \dim \operatorname{cok} P$ depends only on σ_p as a function on $S(\mathfrak{X})$, the unit sphere of $T^*(\mathfrak{X})$ and in fact only on the homotopy class of σ_p. Specifically, let $\pi^*(E_0)$ and $\pi^*(E_1)$ be the vector bundles E_0 and E_1 pulled up to $S(\mathfrak{X})$ via the projection $\pi: S(\mathfrak{X}) \to X$. That is, $\pi^*(E)_s = E_{\pi(s)}$, $s \in S(\mathfrak{X})$). Then $\sigma_P \in C^\infty(\operatorname{Isom}(\pi^*(E_0), \pi^*(E_1)))$. (i) If $\sigma_P = \sigma_{P_1}$, then $\operatorname{ind} P = \operatorname{ind} P_1$. (ii) Any element $\sigma \in C^\infty(\operatorname{Isom}(\pi^*(E_0), \pi^*(E_1))$ equals σ_P for some P. (iii) If σ_0 and σ_1 can be connected by a path σ_t (continuous in the compact open topology), then $\operatorname{ind} \sigma_0 = \operatorname{ind} \sigma_1$. (It is in the verification of (ii) and (iii) that the theory of pseudodifferential operators is particularly useful, even when the original operator P is a differential operator.)

When E is a two step complex, $P: C^\infty(E_0) \to C^\infty(E_1)$ and G is a compact group of endomorphisms, we shall also denote the function $T \to L(T)$, $T \in G$ by $\operatorname{ind}_G P$. In this case σ_P is a G-map i.e. commutes with the action of each $T \in G$ and we write σ_P $\sigma_P \in C^\infty(\operatorname{Isom}_G(\pi^*(E_0), \pi^*(E_1)))$. Because of the discreteness of the characters of the compact group G, and because one can average over G, we obtain d(i)-(iii), with the compact group G added.

(e) Because of (b) and (c), the Hodge theory goes through for a general elliptic complex as in the DeRham complex. That is, let $\Delta = \sum_i (D_i + D_{i-1}^*)^2 = \sum \oplus \Delta_i$. Then $\ker \Delta_i \subseteq C^\infty(E_i)$ and represents $H^i(\mathcal{E})$, i.e. each element of $H^*(\mathcal{E})$ has a unique representative in ker Δ_i. For, if $f \in \ker \Delta_i$, then $D_i f = 0$ and $D_{i-1}^* f = 0$. Thus if $f = D_{i-1} g$, then $D_{i-1}^* D_{i-1} g = 0$ and $f = 0$. Hence the map $\ker \Delta_i \to H^i(\mathcal{E})$ is injective. On the other hand, it is also surjective. For, if $f \in C^\infty(E_i)$ with $D_i f = 0$ and f is perpendicular to ker Δ_i, we can write $f = \Delta_i f_1$ with $f_1 \in C^\infty(E_1)$ by regularity. Hence $f = D_i^* g + D_{i-1} h$ and $0 = D_i f = D_i D_i^* g$ implies $D_i^* g = 0$ so that $f = D_{i-1} h$ and f is a coboundary, i.e., is 0 in $H^i(\mathcal{E})$.

The above shows $H^i(\mathcal{E})$ are finite dimensional. Another application of the Hodge theory is to reduce the computation of $L(I)$ for a general elliptic complex to that of a two-step complex, **i.e. to the** computation of the index of a single elliptic operator P. Namely, $L(I) = \Sigma(-1)^i \dim H^i(\mathcal{E}) = \Sigma(-1)^i \dim \ker \Delta_i$ $= \dim \ker \Delta_+ - \dim \ker \Delta_-$ where $\Delta_+ = \Sigma \oplus \Delta_{2i}$ and $\Delta_- = \Sigma \oplus \Delta_{2i+1}$. However, we have observed that from a general elliptic complex, we can construct the elliptic operator $P = D + D^* : C^\infty(E_+) \to C^\infty(E_-)$. Since $P^* P = \Delta_+$ and $P P^* = \Delta_-$, we conclude that index $D + D^*$ $= L(I)$.

Given the Hodge theory, we can now also give a topological interpretation of the index of $D + D^* : C^\infty(\Lambda_+) \to C^\infty(\Lambda_-)$ of example 3 when k is even. If ω and μ are two closed forms in $C^\infty(\Lambda^{2k})$, then $(\omega, \mu) = \int_M \omega \wedge \mu = \langle \omega, *\mu \rangle$ is a quadratic form which vanishes whenever one is exact. For if $\omega = d\tau$, then $(d\tau, \mu) =$ $\int_M d\tau \wedge \mu = \int_M d(\tau \wedge \mu) = \int_{\partial M} \tau \wedge \mu = 0$. Hence this quadratic form induces a quadratic form on $H^{2k}(M)$. The signature of this quadratic form we call the signature of the manifold M. If we represent $H^{2k}(M)$ by the space of harmonic 2k-forms which is invariant under *, we see that the signature is the dim of harmonic forms in Λ_+^{2k} - dim of harmonic forms in Λ_-^{2k}. But this is just index of $D+D^*$ since ker $D+D^*\wedge\Lambda_+^{i}$ and ker $D+D^*\wedge\Lambda_-^{i}$ have the same dimension for $i \neq 2k$ (using the * operator).

When a compact group G operates, one can choose a Riemannian metrics on X and Hermitian metrics on E_i invariant under the action of G. Hence, when G is a group of endomorphisms of $\mathcal{E}$, G will commute with $D+D^* = P$ and the function $T \to (L(T)$ $T\in G$, becomes the function in the character ring we have denoted by $\operatorname{ind}_G P$.

This reduction does not work when G is not compact, in particular, when a given endomorphism T does not lie in a compact group. This is the essential difference between the fixed point theorm (where assumptions are made concerning the fixed point set of T) and the index theorem (where assumptions are made concerning the group generated by T).

For example, in the case of the **DeRham** complex, any diffeomorphism f induces an endomorphism of the complex, but not necessarily of the induced two step complex C^∞(even forms) $\xrightarrow{d+d^*}$ C^∞(odd forms). If f were an isometry of some Riemannian metrics on M, then f induces an endomorphism on the two step complex. In the case of example 3, since the decomposition of Λ into Λ_+ and Λ_- depends on the metric and orientation f will induce on endomorphism, **if** f is an orientation preserving isometry,

4. <u>The Atiyah-Bott-Lefschetz fixed point formula.</u> We now turn to the statement, examples, and proof of the fixed point formula. We shall discuss the primary case dealing with special endomorphisms called geometric endomorphisms [3]. Namely each T_i is derived from a C^∞ map $f: X \to X$ and a C^∞ bundle map $\varphi_i: f^*(E_i) \to E_i$ where $f^*(E_i)$ is the vector bundle E_i pulled back via f, i.e. $(f^*(E_i))_x = (E_i)_{f(x)}$ and $\varphi_i: (E_i)_{f(x)} \to (E_i)_x$. Then T_i is the map on sections induced by $\varphi_i \circ f^*$.

Note that if x is a fixed point of f, then $\varphi_i(x): (E_i)_x \to (E_i)_x$ is a linear transformation on the vector space $(E_i)_x$, and $(df)(x)$ is a linear transformation on T_x the tangent space of X at x.

<u>Fixed Point Theorem.</u> Let T be a geometric endomorphism of the elliptic complex $\mathcal{E}$ defined by f. Assume f has only simple fixed points (in particular the fixed point set $\mathfrak{F}$ is finite). Then

$$L(T) = \sum_{x \in \mathfrak{F}} \frac{\sum_{i=0}^{N} (-1)^i \operatorname{trace} \varphi_i(x)}{|\det(1-df(x))|}$$

Example 1. In the De Rham complex, with T derived from a C^∞ map $f: X \to X$, the map φ_i is the induced map $df_i: \Lambda^i_{f(x)} \to \Lambda^i_x$. At a fixed point x, some elementary algebra shows that $\Sigma (-1)^i \operatorname{trace} \varphi_i(x) = \det(1-df(x))$. Hence the fixed point formula reduces to the classical Lefschetz formula.

Example 2. Suppose $\mathcal{E}$ is the $\bar\partial$ complex and T is the endomorphism derived from a holomorphic map $f: X \to X$, for X a complex manifold. Since f is holomorphic,

$df\colon T_x \times C \to T_{f(x)} \times C$ preserves the holomorphic and anti-holomorphic tangent spaces; say $df^h\colon H_x \to H_{f(x)}$ and $df^{\bar h}\colon \bar H_x \to \bar H_{f(x)}$. Then the map ϕ_i is the induced map $(df^{\bar h})_i\colon \Lambda^{0,i}_{f(x)} \to \Lambda^{0,i}_x$ and $\Sigma\,(-1)^i$ trace $\phi_i(x) = \det(1-df^{\bar h}(x))$ at a fixed point x. On the other hand, it is easy to see that $|\det(1-df(x))| = \det(1-df^h(x))\,\overline{\det(1-df^h(x))} = \det(1-df^h(x))\,\det(1-df^{\bar h}(x))$. Hence the fixed point formula becomes $L(h) = \sum_{x\in\mathfrak{J}} \det(1-df^h(x))$.

Example 3. Suppose $\mathcal{E}$ is the two-step elliptic complex $d + d^*\colon C^\infty(\Lambda_+) \to C^\infty(\Lambda_-)$ with T induced by an orientation preserving isometry $f\colon X \to X$, X a Riemannian manifold. Then at a simple fixed point x, df is a rotation on the tangent space at x which decomposes into rotations through angle $\theta_j(x)$ on invariant two planes $P_j(x)$. As before the contributions to $L(T)$ cancel except in the middle dimension. Since the ϕ_i are functions of df, the contribution at each fixed point x is a function of the angles $\theta_j(x)$ and some multilinear algebra yields

$$\prod_j \frac{e^{i\theta_j(x)} - e^{-i\theta_j(x)}}{(1-e^{i\theta_j(x)})(1-e^{-i\theta_j(x)})}$$

as the contribution at each fixed point x. Hence the fixed point theorem in this case becomes

$$L(T) = \sum_{x\in\mathfrak{J}} \prod_j \frac{e^{i\theta_j(x)} - e^{-i\theta_j(x)}}{(1-e^{i\theta_j(x)})(1-e^{i\theta_j(x)})}$$

This formula has some interesting consequences. See [3].

5. Proof of the fixed point formula. We have seen that for an elliptic complex $\mathcal{E}$, we can find a parametrix $\{P_i\}_{i=0}^{N-1}$ such that $D_{i-1}P_{i-1} + P_iD_i = I-C_i$ with $P_i : C^\infty(E_{i+1}) \to C^\infty(E_i)$ and C_i of order $-\infty$. Apply the endomorphism T and we find $T_i - T_iC_i = T_iD_{i-1}P_{i-1} + T_iP_iD_i = D_{i-1}T_{i-1}P_{i-1} + T_iP_iD_i$. Since the right side is an endomorphism, so is T_iC_i. Furthermore, T_i and T_iC_i induce the same map on $H^i(\mathcal{E})$ for their difference maps kernel D_i into image D_{i-1}. Hence $L(T) = L(TC)$.

Now T_iC_i is an integral operator with smooth kernel and in fact is of trace class. Suppose $C_i \sim c_i(x,y)$ where $(C_iu)(x) = \int_M c_i(x,y)\, u(y)\, dy$ so that $c_i(x,y) : E_i(y) \to E_i(x)$. Then $T_iC_i \sim \varphi_i\, c_i(f(x),y) : E_i(y) \to E_i(f(x)) \to E_i(x)$, and $\mathrm{tr}\,(T_iC_i) = \int_M \mathrm{tr}(\varphi_i \circ c_i(f(x),x))dx$.

We now show that $L(TC) = \sum_{i=o}^{N} (-1)^i \mathrm{tr}\,(T_iC_i)$. If T_iC_i were of finite rank, this is a standard argument that is the basis of the classical Lefschetz formula, i.e., that the alternating sum is the same on the chain level as on the homology level. We could approximate by operators of finite rank as in Atiyah-Bott. Instead we follow Hormander, using his result which extends the theorem that the trace of an operator of trace class is invariant under change to an equivalent norm on the Hilbert space.

LEMMA. [See 10] Let L_j be trace class operators on Hilbert spaces H_j, $j = 1, 2$. Suppose there exists a closed linear operator S with(i) dense domain in H_1, (ii) dense image in H_2,

(iii) S is one to one, and (iv) $L_2 S \equiv S L_1$. Then $\mathrm{tr}(L_1) = \mathrm{tr}(L_2)$.

We apply the lemma as follows. From the elliptic complex $\mathcal{E}$, pass to the Hilbert space level:

$$0 \to H(E_0) \xrightarrow{D_0} H(E_1) \to \dots \to H(E_{N-1}) \xrightarrow{D_{N-1}} H(E_N) \to 0$$

where D_j are closed operators whose domains lie in $H(E_j)$ and with R_j = range D_{j-1} contained in $N_j = \ker D_j$. Then $T_j C_j = A_j$ are of trace class and $A_{j+1} D_j \subset D_j A_j$, so that $A_j(R_j) \subset R_j$, $A_j(\overline{R}_j) \subset \overline{R}_j$, and $A_j(N_j) \subset N_j$. Hence A_j induces A_j' on $N_j/\overline{R}_j$ which by regularity equals $H^j(\mathcal{E})$. Also A_j induces A_j'' on $H(E_j)/N_j$. It is easy to verify that A_j'' is of trace class and $\mathrm{tr}\,(A_j) = \mathrm{tr}(A_j'') + \mathrm{tr}(A_j') + \mathrm{tr}(A_j|\overline{R}_j)$. Now apply the lemma with $L_1 = A_j''$ $H_1 = H(E_j)/N_j$, $S = D_j : H_1 \to \overline{R}_{j+1} = H_2$ and $L_2 = A_{j+1}|\overline{R}_{j+1}$. We conclude that $\mathrm{tr}\,(A_j'') = \mathrm{tr}(A_{j+1}|\overline{R}_{j+1})$ so that $\mathrm{tr}(A_j) = \mathrm{tr}(A_j') + \mathrm{tr}(A_j|_{\overline{R}_j}) + \mathrm{tr}(A_{j+1}|\overline{R}_{j+1})$. In forming the alternating sum, cancellation gives

$$\Sigma(-1)^j\,\mathrm{tr}(T_j C_j) = \Sigma(-1)^j \mathrm{tr}(A_j) = \Sigma(-1)^j\,\mathrm{tr}(A_j') \text{ so that } L(TC) = L(T).$$

Finally, then, to get the fixed point formula, we must compute $\mathrm{tr}(T_i C_i) = \int_M \mathrm{tr}(\varphi_i \circ c_i(f(x),x))dx$ and by our earlier discussion we can assume that $C_i = C_i(t)$ and $C_i(t) \to I$ boundedly in the strong operator topology as $t \to 0$. Then $L(T) = \lim_{t\to 0} \Sigma(-1)^i \mathrm{tr}(T_i C_i(t))$. Since $C_i(t) \to I$, $c_i(x,y)(t) \to 0$ for $x \neq y$ so that $\lim_{t\to 0} \mathrm{tr}(T_i C_i(t)) = \sum_j \int_{U_j} \mathrm{tr}(\varphi_i \circ c_i(f(x),x))dx$ where U_j is **a coordinate neighborhood of x_j, the jth fixed point. In $U_j x U_j$, write**

$x-y=u$, $x+y=v$ and $c_i(x,y)(t)=k_t(u,v)$. Again $C_i(t) \to I$ will imply that $k_t(u,v)$ will approach the δ-function in the u-variable as $t \to 0$. Thus $c_i(f(x),x)(t) = k_t(x-f(x), x+f(x))$ and

$$\int_{U_j} tr(\varphi_i \circ c_i(f(x),x)(t))dx = \int_{U_j} tr(\varphi_i \circ k_t(u,v)) \frac{du}{|1-df|} \to \frac{tr\, \Phi_i(x_j)}{|1-df|(x_j)}.$$

6. Review of K-theory. The appropriate machinery to express the index theorem is K-theory. In this section we describe this ring and in the next proof the first main theorem, the periodicity theorem.

Let X be a locally compact space. There are several ways of defining K(X) ; the most natural from our point of view is the following. As basic objects we take a complex $\mathcal{F}$ of vector bundles over X : $0 \to F_0 \xrightarrow{\alpha} F_1 \xrightarrow{\alpha} F_2 \dots \xrightarrow{\alpha} F_n \to 0$, with $\alpha^2 = 0$. The support of such a complex is the set of $x \in X$ for which the sequence is not exact, and we deal only with complexes with compact support. Two complexes $\mathcal{F}_0$ and $\mathcal{F}_1$ are homotopic if there exists a complex $\mathcal{G}$ over $X \times I$ such that $\mathcal{F}_i = \mathcal{G}|_{X \times i}$, $i = 0,1$. Under $\oplus$, the set of homotopy classes of complexes forms a semi-group. One obtains the abelian group K(X) by dividing out by the subsemigroup represented by complexes with empty support. The relevant example here is the symbol sequence of an elliptic complex over a compact manifold X , which gives a complex $\tilde{\mathcal{E}}$:

$0 \to \pi^*(E_0) \xrightarrow{\sigma_{D_0}} \pi^*(E_1) \to \dots$ over $T^*(X)$ with support on the 0-section = X . Hence the elliptic complex $\tilde{\mathcal{E}}$ gives an element

$\sigma \in K(TX)$. The analytical properties of the index described in section 3 is easily translatable into the present notation to give a homomorphism a-ind: $K(TX) \to$ integers. The a-ind assigns to each "symbol" in K(TX) the index of the associated elliptic operator.

With no extra difficulty, we can consider complexes with a

compact group G operating, and using maps that commute with the action of G, we get $K_G(X)$ and $a\text{-ind}_G\colon K_G(TX) \to R(G)$, when $R(G)$ is the character ring of G.

The group $K(X)$ can be described in another way. When X is compact, the isomorphism classes of vector bundles over X forms an abelian semi-group under $\oplus$. The associated abelian group is $K(X)$. This amounts to adding the equivalence relation: $E \sim F$ if there exists a vector bundle G such that $E \oplus G \cong F \oplus G$ and then taking formal differences. Any continuous map $f\colon X \to Y$ produces a homomorphism $f^*\colon K(Y) \to K(X)$ coming from pulling back vector bundles from Y to X. In particular, if $x_0 \in X$, we have the inclusion map $j\colon x_0 \to X$ and $j^*\colon K(X) \to K(x_0) \simeq$ integers. The kernel of j^* we denote by $K(X,x_0)$. When X is locally compact, let $X^+ = X \cup \infty$ be its one point compactification Then $K(X) = K(X^+,\infty)$.

We shall not include a proof that the two definitions of $K(X)$ given here are equivalent [1,4]. The main idea goes back to the relation between $K(R^n)$ an the $n-1^{\underline{st}}$ homotopy groups of $Gl(N,C)$ $N \gg n$. Given a map $\sigma\colon S^{n-1} \to Gl(N,C)$, we get a vector bundle E_σ on S^n by writing $S^n = D^+ \cup D^-$, the two discs pasted along their common boundary S^{n-1}. The vector bundle E_σ is obtained by patching $D^+ \times C^n$ to $D^- \times C^n$ along S^{n-1} by σ. Write $S^n = R^n \cup \infty$ where ∞ is the south pole. Then $E_\sigma - S^n \times C^n \in K(S^n,\infty)$ which equals $K(R^n)$ (second definition). It is easy to see that homotopic maps give the same element of

$K(R^n)$. In fact $K(R^n)$ is isomorphic to the homotopy classes of maps of $S^{n-1} \to Gl(N,C)$.

On the other hand, σ can be extended (**radially**) to give a map $\tilde{\sigma}$ of $R^n - 0 \supset S^{n-1}$ into $Gl(N,C)$ so that

$0 \to C^N \xrightarrow{\tilde{\sigma}} C^N \to 0$ is a complex which has compact support (namely the origin) and hence gives an element of $K(R^n)$ (first definition).

Conversely, a vector bundle E on S^n comes from a map σ because $E|D_+ \simeq D_+ \times C^N$ while $E|D_- \cong D_- \times C^N$ so that the map $S^{n-1} \times C^n \to E|_{D_-}|_{S^{n-1}} = E|_{D_+}|_{S^{n-1}} \to S^{n-1} \times C^N$ gives a map $\sigma: S^{n-1} \to Gl(N,C)$ in which $E = E_\sigma$. In this way one establishes the correspondence between the two definitions for $K(R^n)$.

This second definition of $K(X)$, as the vector bundles over X made into a group under $\oplus$ can be motivated analytically when one tries to define the index of a family of Fredholm operators. We shall need this notion in order to give an analytic proof of the Bott periodicity theorem, so we describe it now. See [2],[11].

Let $\mathcal{F}$ denote the space of Fredholm operators on a Hilbert space H , and let $\mathcal{S} : X \to \mathcal{F}$ be a continuous map of the compact space X into $\mathcal{F}$ (operator norm topology on $\mathcal{F}$) . We think of $\mathcal{S}$ as a family of elliptic operators indexed by X . If X is a point x_0 , then $\mathcal{S}$ is a single operator and its index is well defined as an integer or element of $K(x_0)$. In general, we would like to define the index of the family, ind $\mathcal{S}$, when X is compact. If $\{\ker(\mathcal{S}(x))\}_{x \in X}$ were all of the same dimension, this family of vector spaces would form a vector bundle over X

as would the cokernel family. It would be natural, in this case, to define index $\mathcal{J} = \{\ker(\mathcal{J}(x))\}_{x\in X} - \{\operatorname{cok}(\mathcal{J}(x))\}_{x\in X}$. One is forced to consider the formal difference of these two vector bundles, and so obtain an element of $K(X)$.

In general, the family of $\{\ker(\mathcal{J}(x))\}_{x\in X}$, does not form a vector bundle and we must define **index** $\mathcal{J}$ in a slightly more complicated way using the semicontinuity properties of this family. For $x_0 \in X$, choose a finite dimensional subspace $V_0 \subset H$ such that if P_0 is the projection on $V_0^{\perp}$, then the map $P_0 \mathcal{J}(x_0): H \to V_0^{\perp}$ is surjective. It is easy to check that there exists a neighborhood N_0 of x_0 such that for all $x \in N_0$, $P_0 \mathcal{J}(x): H \to V_0^{\perp}$ is surjective and $\{\ker(P_0 \mathcal{J}(x))\}_{x\in N_0}$ is a vector bundle over N_0.

Cover X by finitely many such neighborhoods $\{N_i\}$ with corresponding subspaces $\{V_i\}$. Let $V = \Sigma V_i$ and P be the projection on $V^{\perp}$. It follows that $P\mathcal{J}(x): H \to V$ is surjective for all $x \in X$ and $\{\ker(P\mathcal{J}(x))\}_{x\in X}$ is a vector bundle E over X. Now define **index** $\mathcal{J}$ to be that element of $K(X)$ which is E - the trivial bundle $X \times V$.

In the construction above, if other compliments V'_j were chosen, giving P' equal to the projection on $(\Sigma V'_j)^{\perp}$, then $E - V = E' - V'$ in $K(X)$ by virtue of the equivalence relation $E \sim F$ when $E \oplus G \cong F \oplus G$.

We now turn to some multiplicative properties of $K(X)$. Because of the tensor product operation $\otimes$, $K(X)$ is a ring. The tensor product also gives a map $K(Y) \otimes K(X) \to K(Y \times X)$ as follows: If $\mathcal{E}$ is a complex over X and $\mathcal{F}$ over Y, then their external tensor product $\mathcal{E} \boxtimes \mathcal{F}$ gives one over $Y \times X$ which is compactly supported if $\mathcal{E}$ and $\mathcal{F}$ are. We have observed that for the index we can always reduce to a 2-step complex. In that case, $\mathcal{E} : 0 \to E_0 \xrightarrow{\alpha} E_1 \to 0$ and $\mathcal{F} : 0 \to F_0 \xrightarrow{\beta} F^1 \to 0$. Then

$$\mathcal{E} \otimes \mathcal{F} : 0 \to E_0 \boxtimes F_0 \xrightarrow{\alpha \boxtimes 1 + 1 \boxtimes \beta} E_1 \boxtimes F_0 \times E_0 \boxtimes F^1$$

$\xrightarrow{-1 \boxtimes \beta + \alpha \boxtimes 1} E_1 \boxtimes F_1 \to 0$. Using the adjoint device, this is equivalent to the 2-step complex:

$$0 \to E_0 \boxtimes F_0 \oplus E_1 \boxtimes F_1 \xrightarrow{\gamma} E_1 \boxtimes F_0 \oplus E_1 \boxtimes F_1 \to 0$$ where

$$\gamma = \begin{pmatrix} \alpha \boxtimes 1 & -1 \boxtimes \beta^* \\ 1 \boxtimes \beta & \alpha^* \boxtimes 1 \end{pmatrix}.$$ (One can view the external tensor product as a formal generalization of the De Rhan complex on $Y \times X$ in terms of the De Rhan complex on each factor.)

As a special case, suppose $Y = V$ a complex vector space, so that we have the map $s: K(V) \otimes K(X) \to K(V \times X)$. Now the symbol sequence of the De Rhan complex gives a special element $-\lambda_V \in K(V)$. (The complex is: $1 \to V \to \Lambda^2(V) \to \ldots$ and α is wedge multiplication.) Let $\sigma: K(X) \to K(V \times X)$ be given by $a \longmapsto s(\lambda_V \otimes a)$. More generally, suppose W is a complex vector bundle over X rather than just the product $W = V \times X$. Then $\lambda_W \in K(W)$ and we still have the map $\sigma: K(X) \to K(W)$ obtained by pulling a up to W via the projection map and multiplying by λ_W. The map σ is called the Thom map. When W is a trivial bundle, the Bott periodicity theorem asserts that σ is an isomorphism. In particular, when X is a point, $\sigma: K(pt) \to K(V)$ is an isomorphism.

7. **The Bott Periodicity Theorem. Now let us proof the periodicity Theorem following Atiyah [2]. We note that the multiplicative struc**ture of the previous section holds for $K_G(X)$ as well. However, only some of the proofs of the periodicity theorem generalize to include the K_G case.

Theorem: Let X be locally compact. Then the Thom map $\varphi : K(X) \to K(C^1 \times X)$ is an isomorphism.

Proof: To show it is an isomorphism, we construct a map $\psi_X : K(C^1 \times X) \to K(X)$ which satisfies: (i) if X is a point p, then $\psi_p(\lambda_{C^1}) = 1$. (ii) we have commutativity of the diagram

$$\begin{array}{ccc} K(C^1 \times X) \otimes K(Y) & \longrightarrow & K(C^1 \times X \times Y) \\ \Big\downarrow \psi_X \otimes I & & \Big\downarrow \psi_{X\times Y} \\ K(X) \otimes K(Y) & \longrightarrow & K(X \times Y) \end{array}$$

Given such $\{\psi_X\}$, we show we get an inverse to φ. Apply (ii) with $X = p$. Then for $u \in K(Y)$, $\psi_Y \varphi(u) = \psi_p(\lambda_{C^1}) \cdot u = 1 \cdot u = u$ so that ψ is a left inverse for φ.

Next, let $Y = C^1$ and suppose $u \in K(C^1 \times X)$. Then (ii) gives $\psi_{X \times C^1}(u \lambda_{C_1}) = \psi_X(u) \lambda_{C_1} \in K(X \times C^1)$. Let $\tau : C^1 \times X \times C^1 \to C^1 \times X \times C^1$ which interchanges the first and third factor, so that τ is independent of X and as a real linear transofmation on R^4 has determinant $+1$. Hence τ can be connected to the identity map on $C^1 \times X \times C^1$ and hence $\tau^*(u\lambda_{C_1}) = u\lambda_{C_1}$ in $K(C^1 \times X \times C^1)$. But $\tau = \alpha \circ (1 \times \beta)$ where $\tau : (z_1, x, z_2) \xrightarrow{1\times\beta} (z_1, z_2, x) \xrightarrow{\alpha} (z_2, x, z_1)$. Hence $\tau^* = (1\times\beta)^* \circ \alpha^*$ and $u\lambda_{C_1} = \tau^*(u\lambda_{C_1}) = (1\times\beta)^* \circ \alpha^*(u\lambda_{C_1}) = (1\times\beta)^*(\lambda_{C_1} u) = \lambda_{C_1} \beta^*(u) = \varphi(\beta^*(u))$. Since ψ is a left inverse for φ and we have $\psi_{X\times C^1}(u\lambda_{C_1}) = \psi_X(u)\lambda_{C_1}$, we obtain $\beta^*(u) = \psi_X(u)\lambda_{C_1}$. But β^* reverses order so that finally $u = \lambda_{C_1}\psi_X(u) = \varphi\psi_X(u)$ and ψ is a right inverse for φ.

To construct a map ψ with the desired two properties we use the index of a family of Fredholm operators.

Let $\ell_2 = \left\{ \sum_{-\infty}^{\infty} a_n e^{n\theta}, \ \sum |a_n|^2 < \infty \right\}$ and $H = \left\{ \sum_{n \geq 0} a_n e^{n\theta}; \ \sum |a_n|^2 < \infty \right\}$. Let $P : \ell_2 \to H$ be the projection map. If f is a continuous complex valued function on the circle, then M_f (multiplication by f) is a bounded operator on ℓ_2 and in fact $\|M_f\| = \sup_\theta |f(\theta)|$ and $M_{fg} = M_f M_g$. Let $S_f = PM_f : H \to H$. It is well known that if f is invertible, then S_f is Fredholm and index S_v = minus the winding number of f. (Approximating by trigonometric polynomials shows that $S_{fg} - S_f S_g$ is a compact operator so that $S_{f^{-1}}$ is an inverse to S_f modulo compact operators. The index can be checked on $f = e^{ik\theta}$.)

If V is a fixed n dimensional complex vector space (a vector bundle over a point), let $H_V = H \otimes V$ (n tuples of functions in H) contained in $\ell_2 \otimes V$ and let P_V be projection on H_V. Again if f is a continuous map from S^1 to $G\ell(V)$ (invertible matrix valued function), then M_f is an invertible bounded operator on $\ell_2 \otimes V$ and $S_f = PM_f : H_V \to H_V$ will be Fredholm.

A mild extension of the above is the case where V is a vector bundle over X and $f : S^1 \times X \to \mathrm{Aut}(V)$, i.e., for each $s \in S^1$ and $x \in X$, $f(s,x) = f_x(s) \in G\ell(V_x)$. Now $H \otimes V$ is a Hilbert space vector bundle over X whose fiber at x is $H_{V_x} = H \otimes V_x$. For each $x \in X$, we have the Fredholm operator $\mathcal{S}(x) = S_{f_x}$ on $H \otimes V_x$. In this way, we obtain a family of Fredholm operator $\mathcal{S}^f$ over X, which has an index in $K(X)$. Though our previous discussion of families considered only a fixed Hilbert space, at present we are allowing the Hilbert space $H \otimes V_x$ to vary with x. However, the vector bundle V is locally a product and hence so is $H \otimes V$ and the earlier construction of the index for families carries over.

To construct $\psi : K(C^1 x X) \to K(X)$, we first construct a map $\psi : K(S^2 x X) \to K(X)$. Let E be a vector bundle on $S^2 \times X$ and write $S^2 \times X$ as a union of $D^+ \times X$ and

$D^- \times X$ glued along $S^1 \times X$ where $S^1 = D^+ \cap D^-$ and $D^{\pm}$ are two discs. Since $D^{\pm}$ are homotopically trivial, $E|_{D^{\pm}\times X} \simeq V \times D^{\pm} \times X$ where $V = E|_{1\times X}$, $1 \in S^1$ and V is a vector bundle on X. E is then obtained by gluing $V \times D^{\pm} \times X$ together along $S^1 \times X$ by a map $f : S^1 \times X \to \mathrm{Aut}(V)$, i.e., E is constructed out of the pair (V,f). We now define $\psi(E) = \text{index } \mathcal{L}^f$. It is easily seen to be additive on direct sums of vector bundles and hence gives $\tilde{\psi} : K(S^2 \times X) \to K(X)$. Let $\psi : K(C^1 \times X) \to K(X)$ be the restriction to $K(C^1 \times X)$. (If X is locally compact, use the above for $X \cup \infty$).

Finally, we must check that ψ has the desired properties. Now the element $\lambda_{C^1} \in K(C^1) \subset K(S^2)$ is obtained from the complex $0 \longrightarrow 1 \xrightarrow{\alpha} C^1 \longrightarrow 0$ where at every point $z \in C^1$, α is multiplicative by z. Thus $\lambda_{C^1} = 1-H$ where H is the line bundle obtained from 1 by gluing with $f = e^{i\theta}$. Now index $\mathcal{L}e^{i\theta}$ = minus winding no $e^{i\theta} = -1$ and similarly index $\mathcal{L}^1 = 0$. Hence (i) $\psi_p(\lambda_{C^1}) = \psi_p(1) - \psi_p(H) = 1$.

To verify condition (ii) one merely carries along a vector bundle (over Y) as extra coefficients and the commutativity of the diagram is automatic.

8. <u>The index theorem.</u> We can now describe the topological map $t\text{-ind}_G: K_G(TX) \to K_G(\text{point})$. Let $i: X \to E$ be an equivariant imbedding of X into a vector space E on which G is represented, and let N be a tubular neighborhood of X in E , which can be identified with the normal bundle of X in E . Then TN is a vector bundle over TX whose fiber is two copies of the normal bundle at each point $x \in X$.

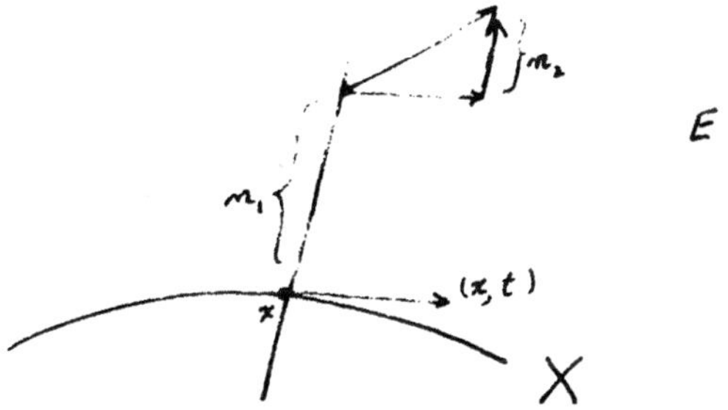

Hence TN is a complex vector bundle over TX where the second copy of N_x is the imaginary part at x . We have then the Thom map $\varphi : K_G(TX) \to K_G(TN)$.

But TN is an open set in TE , i.e. we have the inclusion $k: TN \to TE$ and this induces the map $\tilde{k}^*: K_G(TN) \to K_G(TE)$ (obtained from the map $\tilde{k}: TE^+ \to TN^+$ which collapses $TE^+ - TN$ to ∞ in TN^+). We now define $i_! = \tilde{k}^* \circ \varphi : K_G(TX) \to K_G(TE)$.

When X is a point p and $j: p \to 0 \in E$, then $j_!: K_G(p) \to K_G(TE)$ is just the Thom map of $TE = E \times C$ over a point and hence is an isomorphism. We define $t\text{-ind}_G = (j_!)^{-1} i_!$. One must prove this map is independent of the equivariant imbedding, which is a fairly standard argument.

Index Theorem. $\text{a-ind}_G = \text{t-ind}_G$.

We sketch the main ideas of the proof. Note that we have

$$\begin{array}{c} \overset{\text{t-ind}_G}{\overbrace{\qquad\qquad\qquad\qquad\qquad\qquad\qquad\qquad\qquad\qquad\qquad\qquad}} \\ K_G(TX) \xrightarrow[(i)]{} K_G(TN) \xrightarrow[(ii)]{\tilde{k}^*} K_G(TE) \xrightarrow[(iii)]{(j_!)^{-1}} K_G(p) \ . \\ \underset{\text{a-ind}_G}{\underbrace{\qquad\qquad\qquad\qquad\qquad\qquad\qquad\qquad\qquad\qquad\qquad\qquad}} \end{array}$$

What we must do is chose a symbol in $K_G(TX)$ and a corresponding elliptic operator on X . We must follow the operator and its a-ind_G through the three maps indicated in the diagram. The basic analytic facts about elliptic operators were needed to define a-ind_G . To prove the index theorem we shall need some additional analytic properties.

So let us begin with $\sigma \in K_G(TX)$ and $\sigma = \sigma_A$ where A is a G-invariant elliptic operator $A: C^\infty(E_0) \to C^\infty(E_1)$. We must first see what the Thom map corresponds to analytically.

We want to compare $\text{a-ind}_G\, \sigma$ with $\text{a-ind}_G\, \Phi(\sigma)$. However $\Phi(\sigma) \in K_G(TN)$ and N is not compact so $\text{a-ind}_G(\Phi(\sigma))$ is strictly speaking not defined. However, we have the excision axiom due to Seeley.

Excision: Suppose U is an open G-invariant set in compact G-manifolds X_1 and X_2 , with $k_i: U \to X_i$ $i = 1, 2$, and $\tilde{k}_i^*: K_G(TU) \to K_G(TX_i)$. If $\nu \in K_G(TU)$, then $\text{a-ind}_G \tilde{k}_1^*(\nu) = \text{a-ind}_G \tilde{k}_2^*(\nu)$.

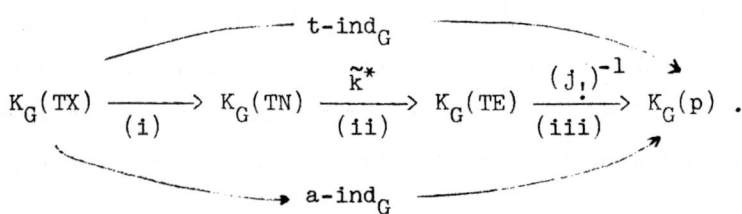

Proof: $\nu \in K_G(TU)$, i.e., ν "vanishes at ∞", means we can choose a compact set $K \subset U$ and vector bundles E and F on U so that $E|_{U-K} = F|_{U-K}$ and $\nu|_{S(U-K)} = I$. Hence $\tilde{k}_1^*(\nu)$ enjoys the same property outside K. If P is an elliptic operator on X_1 of order 0 with $\sigma_P = \tilde{k}_1^*(\nu)$, then $Q = fPf + (1-f^2)I$ with $\operatorname{supp} f \subset U$ and $f \equiv 1$ on K also has $\sigma_Q = \sigma_P$. But if $v \in \ker Q$, then $\operatorname{supp} v \subset U$. Similarly for $\ker Q^*$. Since Q is well defined on X_2 and its symbol is $\tilde{k}_2^*(\nu)$, we have $\text{a-ind}_G Q$ is the same in X_1 and X_2. The excision axiom is proved and we can compute $\text{a-ind}_G(\varphi(\sigma))$ by imbedding N as an open set in any compact manifold.

We use this axiom as follows. Given $\varphi(\sigma) \in K_G(TN)$ instead of imbedding $N \subset E$ (that is in $E \cup \infty = S^m$), we imbed N in a sphere bundle $\tilde{N}$ over X, $k_1 \colon N \to \tilde{N}$ and verify that $\text{a-ind}_G\, \tilde{k}_1^*(\varphi(\sigma)) = \text{a-ind}_G \sigma$.

Let $\tilde{N}$ be the double of $\bar{N}$, i.e., two copies of $\bar{N}$ pasted along $\partial\bar{N}$ = normed sphere bundle. Thus $\tilde{N}$ is a sphere bundle over X and a G-space. Now from the definition of the Thom map, $\varphi(\sigma) \in K_G(TN)$ is $\lambda_{TN}\cdot\sigma$, i.e., σ pulled up to TN and multiplied $\lambda = \lambda_{TN}$. We have interpreted this multiplication earlier as the symbol

$$\lambda \# \sigma = \begin{pmatrix} \sigma \boxtimes I & -I \boxtimes \lambda^* \\ I \boxtimes \lambda^* & G^* \boxtimes I \end{pmatrix}$$

which is an elliptic symbol on the sphere bundle $\tilde{N}$ over X. To compute the a-ind_G of this symbol we need the

<u>Multiplicative Axiom</u>. Suppose (i) Y is a fibre bundle over X with fibre Z and group H; (ii) X and Y are G-spaces; (iii) A is an elliptic operator of order 1 on X commuting with the action of G; (iv) B is an elliptic operator of order 1 on Z commuting with the action of $G \times H$ and $\text{a-ind}_{G\times H}(B) = 1 \in R(G \times H)$. Then $\sigma_B \# \sigma_A \subset K_G(TY)$ and $\text{ind}_G\, \sigma_B \# \sigma_A = \text{ind}_G\, \sigma_A$.

This axiom is a generalization of the product space case [15] and the proof is too technical to give here [4]. It should be added however that in the bundle case, the multiplicative property does not hold in general. It is essential that $\text{a-ind}_{G\times H}(B)$ be a constant in $R(G \times H)$.

Granted this axiom, we can conclude that $\text{a-ind}_G\, \sigma$ $\text{a-ind}_G\, \varphi(\sigma)$ with $\varphi(\sigma) \in K_G(\tilde{N})$ provided we can show that the operator B with symbol λ_0 on the sphere S^n has $\text{a-ind}_{G\times O(n)} = 1$.

Let us look at this normalization axiom in more detail.

The fiber of $\tilde{N}$ is a sphere $S^n = B^n_+ \cup B^n_-$, the union along $S^{n-1} \subset \partial B^n_\pm$ of two copies of the unit ball in R^n. Then $TS^n = B^n_+ \times R^n \cup B^n_- \times R^n$ pasted along $S^{n-1} \times R^n$ by $(x,v) \longmapsto (x,h_x v)$ where h_x is reflection in the hyperplane of R^n perpendicular to x. Consider the complex $\mathcal{J}$ of vector bundles over $OB^n_+ \times R^n = T(OB^n_+)$ (where OB^n_+ is the open ball) with $OB^n_+ \times R^n \times \Lambda^i(C^n) \to OB^n_+ \times R^n \times \Lambda^{i+1}(C^n)$

given by $(x,v,w) \longmapsto (x,v,(v-ix) \wedge w)$. This is exact outside $x = 0$ so this complex defines an element $\mu_+ \in K(T(OB^n_+))$. Since $T(OB^n_+)$ is open in $T(S^n)$, as earlier we have the map $\tilde{k}^*: K(T(OB^n_+)) \to K(TS^n))$. Let $\tau_+ = \tilde{k}^*(\mu_+)$.

The construction of τ_+ is $O(n)$-invariant, so in fact $\tau_+ \in K_{O(n)}(TS^n)$. Furthermore the complex $\mathcal{F}$ is part of a family of complexes $\mathcal{F}_s$ on TS^n where $B^n_{\pm} \times R^n \times \Lambda^i(C^n) \to B^n_{\pm} \times R^n \times \Lambda^{i+1}(C^n)$ given by $(x,v,w) \to (x,v,(v-isx) \wedge w)$. When $s = 0$, $\mathcal{F}_0$ is just the De Rham symbol sequence. When n is even, using the fact that $\text{a-ind}_{O(n)}$ of the De Rham sequence is $2 \in R(O(u))$, together with symmetry considerations of reflection about the equator S^{n-1} gives $\text{a-ind}_{O(n)} \tau_+ = 1 \in R(O(u))$. When n is odd, the same result is true but the argument is a little more complicated.

Comparing the definition of μ_+ with the basic Thom element λ_0 easily shows that $\lambda_0 = \mu_+$. Furthermore everything commutes with the action of G , so we finally have $\text{a-ind}_G\, \sigma = \text{a-ind}_G\, \phi(\sigma)$. On the other hand, using the excision axiom $\text{a-ind}_G\, \phi(\sigma) = \text{a-ind}_G \tilde{k}^*\, \phi(\sigma)$ so that $\text{a-ind}_G\, \sigma = \text{a-ind}\ i_!(\sigma)$. Applying this result to $j_!$ gives $\text{a-ind}_G\, \sigma = \text{a-ind}\ j_!^{-1} i_!(\sigma) = \text{a-ind}\ (\text{t-ind}\ (\sigma))$. But with X a point p and $r \in K_G(p) = R(G)$, $\text{a-ind}\ r = r$ so that finally $\text{a-ind}_G\, \sigma = \text{t-ind}_G\, \sigma$, completing our sketch of the proof of the index theorem.

9. Some applications. When $G = (e)$, the t-ind is computable using the theory of characteristic classes. We shall not enter into that theory, except to say that applying the index theorem to the examples in section 2 gives Chern-Gauss-Bonnet, Hirzebruch, Riemann-Roch, Hirzebruch signature, and integrality theorems.

Of more interest to analysts is what the theorem implies for trivial bundles, i.e., systems. Some of the results are:

(a) Let X be a compact manifold of dimension n in R^{n+1} and suppose P is an elliptic pseudodifferential operator on N-tuples so that σ_p: $S(X) \to G\ell(N,C)$. If $N < n$, then ind $P = 0$. If $N = n$, let $r: G\ell(N,C) \to S^{2n-1}$ which assigns to any nonsingular $n \times n$ matrix its first column normalized to have length 1. Then ind $P = -\frac{1}{(n-1)!}$ degree $(r \circ \sigma)$. If $N > n$, then σ is homotopic to a map $\sigma_1: S(x) \to G\ell(n,C) \subset G\ell(N,C)$.

(b) For any X and with $N \leq n/2$, then ind $P = 0$ unless $n - 2N$ is divisible by 4 and the Euler characteristic of X is zero.

(c) If n is odd and P is a differential operator, then ind $P = 0$. This result does not depend on the index formula, only on the fact that $\sigma_p \in K(TX) \otimes R$ is zero. It can be obtained more directly [15].

When G is not trivial, the results in [4] allow one to compute the character a-ind$_G(P)$ on $g \in G$ in terms of the index of elliptic operators on the fixed point set of g, a kind of Lefschetz formula, and giving a characteristic class type formula for a-ind$_G$. Though this formula is formidable

[5], it is computable when applied to the standard examples. A simple case is this. Let X be a connected complex 2-manifold and T a nontrivial holomorphic involution so that $G = \{I, T\}$. Suppose the fixed point set of T consists of N isolated points and M complex irreducible curves $\{D_h\}_{h=1}^{M}$. Then for the $\bar{\partial}$-elliptic complex,

$$L(T) = N/4 + \sum_{k=1}^{M} \left(\frac{1 - \text{genus}(D_k)}{2} + \frac{D_k^2}{4}\right)$$ where D_k^2 denotes the self intersection. Note that if T has only isolated fixed points, then $L(T) = N/4$, agreeing with the generalized Lefschetz formula when T is transversal, for $df = -I$ at each fixed point.

Bibliography

1. M. F. Atiyah, K-theory, Benjamin, 1967

2. M. F. Atiyah, Bott Periodicity and the Index of Elliptic Operators, Quart. J. of Math. 19 (1968), 173-140.

3. M. F. Atiyah and R. Bott, A Lefschetz Fixed Point Formula for Elliptic Complexes, I, Ann. of Math., 86 (1967), 374-407

4. M. F. Atiyah and G. B. Segal, The Index of Elliptic Operators II, Ann. of Math., 87 (1968), 531-545

5. M. F. Atiyah and I. M. Singer, The Index of Elliptic Operators I and III, Ann. of Math., 87 (1968), 484-530 and 546-604

6. M. F. Atiyah and I. M. Singer, The Index of Elliptic Operators on Compact Manifolds, Bull. Amer. Math. Soc., 69 (1963) 422-433

7. A. P. Calderon and A. Zygmund, Singular Integral Operators and Differential Equations, Amer. J. of Math., 79 (1957), 901-921

8. F. Hirzebruch, Topological Methods in Algebraic Geometry, November, 1966

9. L. Hormander, Pseudo-differential Operators, Comm. Pure Appl. Math., 18 (1965), 501-517

10. L. Hormander, A remark on operators of trace class (to appear)

11. K. Jänich, Vektoraumbundel und der Raum der Fredholm Operators, Math. Ann., 161 (1965), 129-142

12. M. Karaubi, Cohomologie des categories de Banach, C. R. Acad. Sci., (Paris) Sér. A-B, 263 (1966), A275-A278

13. J. J. Kohn and L. Nirenberg, An Algebra of Pseudo-differential Operators, Comm. Pure Appl. Math., 18 (1965), 269-305

14. R. Palais, Seminar on the Atiyah-Singer Index Theorem, Ann. of Math. Study 57, Princeton, 1965

15. R. T. Seeley, Integro-differential Operators on Vector Bundles, Trans. Amer. Math. Soc., 117 (1965), 167-204

16. R. T. Seeley, The Powers A^s of an elliptic operator A, Proc. Symp. on Sing. Int., Chicago, 1966

17. G. B. Segal, Equivariant K-Theory, Publ. Math. Inst. Hautes Etudes Sci., Paris, 1968

M.I.T., Cambridge, Mass.

Stampa: Editoriale Grafica - Roma - Tel. 5890154

GPSR Compliance

The European Union's (EU) General Product Safety Regulation (GPSR) is a set of rules that requires consumer products to be safe and our obligations to ensure this.

If you have any concerns about our products, you can contact us on ProductSafety@springernature.com

In case Publisher is established outside the EU, the EU authorized representative is:

Springer Nature Customer Service Center GmbH
Europaplatz 3
69115 Heidelberg, Germany

Batch number: 09490862

Printed by Printforce, the Netherlands